T0091888

Travels with Curiosity

Charles J. Byrne

Travels with Curiosity

Exploring Mars by Rover

Charles J. Byrne
Image Again
Middletown, NJ, USA

ISBN 978-3-030-53804-0   ISBN 978-3-030-53805-7 (eBook)
https://doi.org/10.1007/978-3-030-53805-7

Cover image: In this image, Curiosity is parked at the foot of Mount Remarkable in the Kimberly site; her MastCam is examining a rock that has tumbled from the hill. The surface of Mars is covered with fine red dust that also floats in the atmosphere. The color images are often filtered like this to show contrast in the rocks for the geologists. That turns the dusty sky to light blue. Credit: NASA

This Springer imprint is published by the registered company Springer Nature Switzerland AG
The registered company address is: Gewerbestrasse 11, 6330 Cham, Switzerland

Preface

The Mars Science Laboratory and the Curiosity Rover

The project Mars Science Laboratory (MSL) was so named to emphasize its objective to land a large mobile array of instruments on the surface of Mars. These instruments were to be designed by scientists in diverse disciplines to gather information not only about the surface as it is today but also how it came to be that way in the course of its history.

The centerpiece of the MSL project is the rover, large enough and strong enough to carry an entire laboratory of instruments and support equipment as well as travel over a diversity of terrain conditions while gathering data for scientific investigation of Mars.

Scientific Curiosity

The winner: Clara Ma

Clara Ma was a 12-year-old student at the Sunflower Elementary School in Lenexa, Kansas when she won NASA's name contest. Asked why she entered the contest, she replied:

"I was really interested in space. But I thought space was something I could only read about in books and look at during the night from so far away. I thought that I would never be able to get close to it, so for me, naming the Mars rover would at least be one step closer."

Scientists, of course, are as diverse in their characters as any other groups of people. But there is one common bond that draws them together in the activity of science. That is a desire to learn more about nature—the Earth, the biosphere, the Solar System, and the universe. In short, how things work. A word for this common bond is CURIOSITY. And that is the name Clara Ma chose as the title for her winning essay submitted to NASA's rover name contest.

Clara signed her name on the rover while it was in test; the rover passed through space on its way to Mars and has been on the surface there for seven years now. A line from Clara's essay is:

> We have become explorers and scientists with our need to ask questions and to wonder.

Some of the questions that the MSL Science Team ponders are: Since life is so pervasive on Earth, how come there is so little evidence of life on Mars? Is there any evidence that there once was life on Mars? Could there have been? To address these questions, the Curiosity rover has instruments that can investigate the habitability of Mars. By examining the current geology of Mars, the nature of the past environments that formed them can be inferred, as can be the habitability of those environments. Curiosity's instruments are discussed in Chap. 2.

Fig. 1 Test models of three JPL designs of Mars rovers are shown with two JPL engineers. Sojourner (bottom left) was landed in 1997, and Spirit and Opportunity (above Sojourner) landed in 2004. Curiosity (right) was landed in 2012. (Image courtesy of NASA, JPL-Caltech.)

The MSL Science Team: Each of the 10 science instruments was proposed by principal investigators. These senior planetary scientists also appointed members of their staff, to be members of the MSL Science Team, a group of about 160 scientists who are responsible, along with their collaborators, for the tactical decisions of Curiosity's activities.

These scientists, about 400 in total, are listed in Appendix A. The management, performance, and productivity of this group of scientists, representing several relevant disciplines and many academic institutions, are discussed in Chap. 11.

Curiosity is the heaviest rover that has ever been landed on any planet but Earth, weighing in at about 2 tons (see Fig. 1). The Apollo Lunar Modules, which landed humans on the Moon, had a dry weight of about 6 tons.

In addition to the construction of the Curiosity rover, the MSL project included figuring out the means of delivering such a heavy load safely to Mars, using a soft landing near a designated site. Landing from space to the surface of Mars is actually more of a challenge than landing on Earth or the Moon. Its thin atmosphere is both a help in dissipating the vehicle's energy and a hindrance because of high winds.

Other parts of the MSL are the communications, propulsion, and support systems used in the transfer trajectory from Earth. The aeroshell (see Fig. 2) contains the ablative shield used in early entry to the Mars atmosphere and a parachute to further reduce the speed. Retro rockets and a radar system maneuver Curiosity to hover over the landing site after it deploys its wheels. Finally, a sky crane gently lowers

Fig. 2 The MSL, assembled for testing. Curiosity is inside the aeroshell (above the black heat shield). (Image courtesy of NASA, JPL-Caltech.)

Curiosity to a landing, ready to roll. The MSL components that support the entry, descent, and landing are discussed in Chap. 3. Someday, similar components scaled by payload weight, may bring astronauts, scientists, and (who knows?) tourists safely to the surface of Mars.

The Tour

What would it be like to travel on Mars? For starters, it would be different from anything you could experience on Earth. First, you would be conscious of not only the spacesuit you are wearing to protect you from the low-density Mars atmosphere but also the equipment in your backpack that would provide the air you breathe. Yet, although heavy on Earth (it would weigh 50 pounds), it only would weigh a sixth of that on Mars.

The environment of Mars itself would be disturbing. First, the sky is red (if the Sun is shining) because of the pervasive fine dust in the atmosphere. Here on Earth, the moisture in the atmosphere scatters sunlight and makes the color of the sky blue if there are no clouds. On Mars, the tiny particles of red dust (iron oxide) do something similar. The red light also causes the color of everything you see to be tinted reddish, so your suit, if white, would appear to be pink.

Many of the images in this book came from the navigation cameras, which are grayscale, but there are many color pictures as well from the science cameras. Some of these are in the natural reddish tints of Mars, but others are color-shifted so that minerals that are similar to those on Earth look familiar. You can tell that these images have been color-shifted because the sky is blue!

Time does not fly, it crawls on Mars!

Your timepiece would be similar to one for Earth, but the duration of a day would be a little different. Because Mars rotates slower than Earth, about 40 minutes each Earth day, the time between sunrise and sunrise would be longer than a day. To avoid confusion, we do not use "day" for this important interval: the term on Mars is "sol." For planning, you would divide a sol into 24 hours and each hour into 60 minutes, but you would be aware that each Mars interval would be slightly longer than a corresponding Earth interval. You should remind yourself of the difference by saying yestersol and tosol and tomorrowsol for yesterday and today and tomorrow, especially when talking to a person on Earth. Actually, you would probably be texting, because of the long minutes (about seven each way) of delay time:

"How are you?"……(fourteen minutes later)……"Fine!"

As you look around you, there would be familiar landforms such as sand, rocks, hills, valleys, but no vegetation whatever. Even though you are high on a hill or ridge with the horizon many kilometers away, there is not a trace of a green tint or water, either! Few places on Earth are so barren.

Still, as we follow the Curiosity rover's trail of exploration together, we would see many signs of powerful flows of water in ancient times having eroded the landscape much as it does on Earth. Such ancient flows have been a determinant of mineral composition of the rocks because the presence of water influences how atoms combine to form molecules and crystals of rocks.

The Curiosity rover will be your guide as she explores the floor of Gale Crater and ascends Mount Sharp. Its many cameras will provide the views as we progress, including the deployment of its instruments. We will provide comments on what has been planned on the ground for Curiosity to perform. Also, we will summarize the interpretation of the data on minerals and structural formations by the diverse scientists as they publish their results. You will be able to join this exciting, ongoing exploration not merely by a robot but also by scores of scientists paving the way back on Earth.

Background of the Mars Science Laboratory

The Jet Propulsion Laboratory, a division of Caltech, came to NASA from the DOD when NASA was founded. Its early history was focused on the Moon (Ranger, Surveyor) but later, as rocket science matured, managed larger, more ambitious projects: Viking landed on Mars, Voyager 1 and 2 visited several planets in fly-bys before they left the Solar System. Mariner flew by Mars. NASA and JPL missions explored Venus, Mercury, Saturn, and Jupiter.

A new NASA administrator, Daniel Goldin, was appointed in 1992. He introduced a new initiative, FBC, standing for "Faster, Better, Cheaper." He was responding to a period when NASA's programs were overrunning their budget and running late. In some ways, there were successes in this new management style, but in others there were failures.

In the particular case of JPL, an early success after NASA endorsed the FBC policy was the Pathfinder Program, which landed the first roving vehicle, Sojourner, on Mars at Ares Vallis on December 4, 1996. Pathfinder, managed by Tony Spear, was often seen as a validation of the FBC policy. Sojourner, managed by Donna Shirley, included an X-ray diffraction instrument similar to one carried by Curiosity. Together, Pathfinder and Sojourner were light enough for an airbag landing. After their safe landing and deployment, a number of rocks and soil samples were analyzed, finding both similarities and differences to Earth rocks. Power was supplied by solar panels to supplement a battery. Sojourner managed by Donna Shirley operated for 3 months and drove a total of 100 meters on its six wheels, a configuration that was successfully used for Spirit, Opportunity, and Curiosity. It also will be used for Mars 2020.

Sojourner's success as an FBC project can be attributed to limited and well-chosen objectives, taking advantage of previously established technologies, yet also taking focused development risks. Unfortunately, the next two JPL missions, Mars Polar Lander, launched January 3, 1999, and Mars Climate Orbiter, launched December 11, 1998, were unsuccessful.

Was JPL trying to do too much? Or were the faster and cheaper rules getting in the way of success? NASA and JPL changed strategy after reviewing the success and failures of the FBC policy.

NASA asked the Academy of Sciences to survey the national and global community of planetary scientists on which of their goals was most important in the decade of 2003–2013. This was a successful step toward producing missions that would be individually productive, mutually supportive, and affordable. It was the first planetary decadal survey, published in 1999. The second decadal survey, published in 2013, was for the years 2013–2022.

This first decadal survey was entitled "New Frontiers in the Solar System. An Integrated Exploration Strategy." The duration of the study period was chosen to be 10 years to allow for a reasonable time to develop proposals, engineer, develop, and test the spacecraft and obtain the launch vehicles.

The First Decadal Survey

New Frontiers in the Solar System: An Integrated Exploration Strategy

National Research Council 2003. Washington, DC: The National Academies Press.

https://doi.org/10.17226/10432. Springer, 1999.

Chapter 3 of the document on Mars objectives is entitled "The Evolution of an Earth-like Planet." Exploration of Mars was given a high priority: over 90% of the responses put it as one of their top five targets for the period of the study. The next two were the Moon (65%) and Europa (62%). The discussion points out its similarity to Earth as a rocky planet of similar size, having an atmosphere, and apparently once having had liquid water on its surface. Was it ever habitable? If so, did it have life? Does it have life now, perhaps below the surface?

NASA and JPL would adopt a long-range, ambitious goal "to look for signs of life on Mars." That goal was given credence by evidence, provided by Sojourner and other missions, of an early Mars climate that would allow liquid water on the surface. The new strategy was to pursue that ambitious goal with a series of missions, each with limited subordinate goals, and also to limit risks through redundancy. Two orbiters and two rovers were planned. The rovers were about the same mass as the Pathfinder and Sojourner together, small and light enough for an airbag landing.

In this new strategy, an added goal was to support future human landings on Mars. For this purpose, one of the orbiters was named Mars Reconnaissance Orbiter and would have a high-resolution camera, to play a similar role that Lunar Orbiter played for the Apollo Program—landing site selection and certification. These steps gained support for the new robotic program as did the prospect of increased longevity and mobility on the Mars surface.

The experiences of the FBC era were settled for JPL with the new goals. JPL now felt that it could take on the challenge of searching for signs of early life on Mars. An important subordinate task would be to mobilize the scientific community to rise to the challenge. How do you determine what is or is not a sign of life? This was found to be too difficult a task in the first decadal period. Instead, the goal would be to first try to determine if the Mars environment was ever habitable in the first place.

The new project was named the Mars Science Laboratory (MSL). While mobility was essential in its mission, as learned in the Apollo Program when it added a roving vehicle (for two astronauts!), it would be a laboratory on wheels, not just a vehicle with instruments.

Other requirements, not only supportive of the science mission but also supportive of a manned mission, would be extended lifetime and range (relative to Sojourner's 3 months and 100 meters) and improved landing precision. Payload weight had to increase. Airbags would not scale to that problem (and explorers would not miss bouncing and rolling to a stop), so a new plan for entry, descent, and landing would be needed.

The landing site selection could be influenced, in the course of development, by information gathered from the fleet of two orbiters, 2001 Mars Odyssey and Mars Reconnaissance Orbiter, and two mobile landers, Spirit and Opportunity. Such flexibility required additional navigational capability, especially in latitude and its attendant thermal range.

A particularly difficult challenge was to develop a new technology to take samples of rocks and analyze them for not only their element composition but also their molecular and mineral structure, in order to determine environmental conditions during the rock's formation and modification. The specific goals for the Mars Science Laboratory would not be the detection of life (or past life) on Mars but rather to try to answer the questions: Was there ever an environment on Mars that could support life (as we know it)? How long could that environment have existed?

As the proposal for the MSL was being developed, the launch window was set for 2009. However, as the implications of the requirements were realized, the delivery of instruments and the heat shield were late for testing. Engineering changes needed as tests failed caused further delays. A much more detailed discussion of the problems of this period can be found in *The Design and Engineering of Curiosity* by Emily Lakdawalla, Springer, 2018.

The result of these problems and others was that the launch window had to be delayed to 2011, at a large hit to the NASA budget. The cost was absorbed in the unmanned program's budget, generating some hard feelings in other programs. The two-year slip on such a large project put JPL management and workers on a lot of pressure to survive in a very difficult mission.

Fortunately, the four JPL Mars missions that were planned for the prelude to the Mars Science Laboratory were successful. 2001 Mars Odyssey launched on April 7, 2001 and entered orbit on October 24, 2001. Mars Reconnaissance Orbiter launched on August 12, 2005 and entered orbit on March 10, 2006. Spirit launched on June 10, 2003 and landed on January 4, 2004, operating for six years before becoming stuck in deep sand. Opportunity launched on July 7, 2003 and landed on January 25, 2004. It was still in operation, exploring Endeavor crater, when MSL rover Curiosity landed and continued for a record of 5,250 Earth days for rover endurance on Mars! Meanwhile, the orbiters are still acting as high-speed data relays and tactical camera support for Curiosity.

Middletown, NJ, USA Charles J. Byrne

Acknowledgements

My interest was diverted from the Moon to Mars (that is quite a distance) by my son, Daniel Joseph (DJ) Byrnc. DJ has had a long career at JPL; he wrote and tested the radar pre-processing program for the powered descent phase of the Mars Science Laboratory. Guided by the radar, the sky crane dropped Curiosity well inside its target ellipse. He is now working on Mars 2020. He contributed the last chapter of this book, describing the exciting improvements in engineering and science capability of this next NASA mission to Mars. Mars 2020 is scheduled to launch in July 2020 and land in Jezero crater on February 18, 2021, at 20:30 GMT.

In the course of writing this work, I am indebted first to NASA for their policy of rapid release of raw data and timely news releases. The images from the engineering cameras and the science cameras are posted on web pages within days, sustaining your interest (what did Curiosity see today).

Then there are the journalists, especially the regular bloggers like Emily Lakdawalla, of the Planetary Society. In addition to her blogs, her book "The Design and Engineering of Curiosity," was very helpful. I am aware of how much research Emily has done because I prepared a talk on the topic for my local astronomy group in the summer of 2018, before her book was available, and had to plunge into the diverse engineering sources to find out how the rover worked. My work got much easier when I got a copy of her book and then harder again when the book's coverage of the early surface operations ended in March of 2017.

Two other journalists, Phil Stooke of the Planetary Society and Robert Burnham of Arizona State University have been a very useful resource. Phil provides overlays for the track maps regularly, annotating the stops of Curiosity every time it moves, labeling the stop names. Robert edited "The Red Planet Report" for ASU's Mars Space Flight Facility until recently, providing reports from members of the nearly daily Curiosity operations meetings.

I am deeply indebted to my patient proofreaders, first my diligent, loved, and loving wife, Mary R. Byrne, who corrected the early drafts. DJ Byrne brought his JPL and Mars experience to bear. Then the Springer editor, Maury Solomon now retired, and her successor, Hannah Kaufman, who have now brought my fifth book toward publication.

Finally, I honor the many people all over the world, including those from Caltech, JPL, and the many Science Team Members, engineers, and technicians from many other academic institutions and corporations as we enter the shadow of the coronavirus pandemic. Stay safe and carry on teleworking and following safe practices in essential activities: learning is an essential activity.

Operations of the rovers, Curiosity now at Mars and Perseverance near the launch site, are or will be carried on by teleworking, with the exception of final assembly in a clean room and Earth-based testing of work-arounds for unexpected events, which will be done, as all kinds of collective activities, following safe practices.

Specific information for each project is in Chaps. 11 and 12. The status of each of these missions can be followed by going to mars.nasa.gov or searching for Mars Exploration Program.

Contents

1 The Mars Science Laboratory Mission

The Mars Science Laboratory mission is derived from NASA's strategic vision for Mars, called the Mars Exploration Program. It is the product of the consultation by NASA scientists with American and international scientists who are active in the analysis and design of observational instruments that supplement the results of previous missions.

Program Goals

The Mars Exploration Program set forth four goals:

Goal I: *Determine If Mars Ever Supported Life*
Goal II: *Understand the Processes and History of Climate on Mars*
Goal III: *Understand the Origin and Evolution of Mars as a Geological System*
Goal IV: *Prepare for Human Exploration*

(From MEPAG, 2018, Mars Scientific Goals, Objectives, Investigations, and Priorities. D. Banfield, ed., 81 pp. white paper posted October 2018 by the Mars Exploration Program.)

NASA chose the name "Mars Science Laboratory" to emphasize that the previous spacecraft (landers, rovers, fly-bys, and orbiters) were directed toward relatively specific goals. Earlier missions have gathered a considerable understanding of Mars and that planet's relation to Earth, but many new questions have arisen.

Sets of overlapping disciplines are needed to come to an understanding of not only what is the nature of the surface and atmosphere today but how did it come about? What processes produced the geologic formations we see now? Dr. John Grotzinger, now Chief of the Division of Geological and Planetary Sciences, Caltech, was the first Chief Scientist, who led teams of about 400 researchers working in many institutions. Later Dr. Aswin Vasavada, initially MSL Project Scientist, JPL, replaced Dr. Grotzinger.

A difficult question is whether conditions were ever favorable for life, and if so was it a long enough time to allow life to emerge and leave evidence we could find today. Since Mars is a rocky planet much like Earth (but with important differences), it is a promising place to search for life. On Earth, the time interval between the emergence of life and its evolution to the stage when it left evidence of its existence extends from when the surface had cooled sufficiently for water to be liquid to the time when generations of bacteria could form the massive stromatolites (organized mats of bacteria), the earliest fossils found so far. The interval between these two times is estimated to be about 700 million years. Of course, that period is a single data point, neither a maximum nor a minimum period for life to begin and also to leave evidence to be detected in our time.

C. J. Byrne, *Travels with Curiosity*, https://doi.org/10.1007/978-3-030-53805-7_1

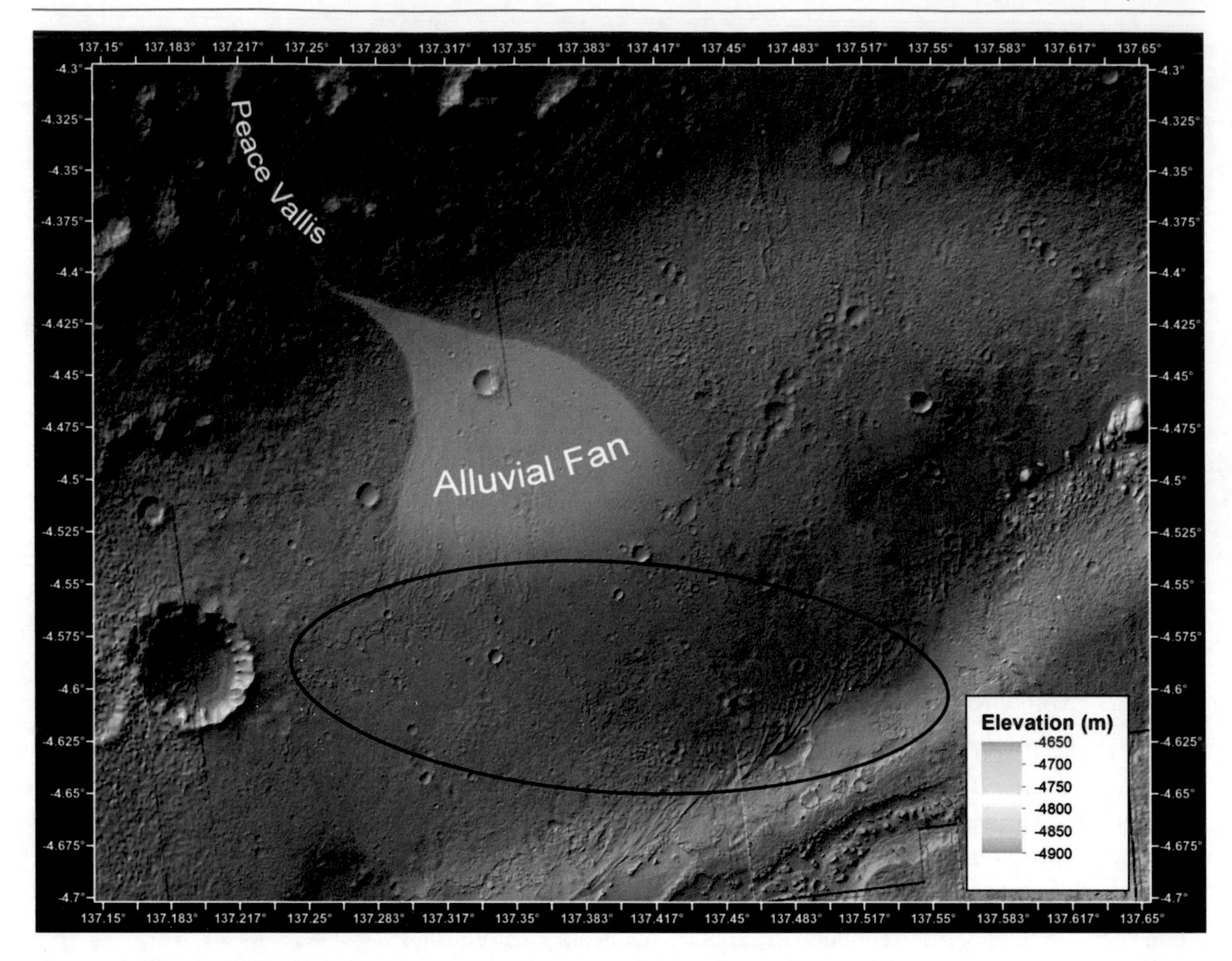

Fig. 1.1 This false-color topographic map displays elevation, derived from orbital observations. It shows the northern edge of Gale crater and Peace Valley (officially Peace Vallis). The ellipse is the landing target, and a small black + marks the landing site of Curiosity. Water from the surrounding region ran down Peace Vallis, spreading out into channels in the alluvial fan. (Image courtesy of NASA/JPL-Caltech)

Since life as we know it is dependent on water, the history of water on Mars is critical to the search for evidence of life there. So the role of water in the geology and geochemistry of Mars are critical fields of study. This strategy is sometimes called "Follow the water." One of many examples of erosion by fluid flow is shown in Fig. 1.1. Curiosity has been able to confirm that the fluid that eroded Peace Valley was water.

Goals II and III make it clear that investigation of the atmosphere and geology of Mars are objectives in themselves. They not only increase our knowledge of Mars and Earth but also (by inference) of the other rocky planets of the Solar System and exoplanets. In relation to Goal III there is a small volcanic area near the top of Mount Sharp. Instruments can sense data up to a few centimeters below the surface as well as from cliffs and other features presented by the terrain. Examples of geological formations are shown in Fig. 1.2. These formations, which come together in this single image, are typical of those seen by Curiosity throughout its travels on lower Mount Sharp.

Fig. 1.2 Murray formation (center) and Stimson formation (upper left), with windblown sand. (Image courtesy of NASA/JPL-Caltech/MSSS, from a panorama cropped by the author)

Fig. 1.3 The Curiosity rover photographed this mosaic of clouds on May 17, 2019, with a Navcam. (Image courtesy of NASA/JPL-Caltech)

Goal II, concerning the atmosphere, is addressed by two suites of instruments: the Sample Analysis at Mars (SAM), which periodically sniffs and analyzes the local atmosphere, and the Rover Environmental Monitoring System (REMS), which collects data such as temperature, pressure, and humidity. Also, the camera systems periodically scan the sky for clouds, dust, opacity, and dust devils (Fig. 1.3).

Goal IV addresses the question of whether humans can explore, colonize, or gain resources from Mars. As humans, we have adapted to nearly all the environments on Earth, even modifying our genetics (by evolution) to do so. Answers to these questions come from measuring and monitoring variations in the Martian surface, subsurface, and atmosphere. The MSL entry, descent and landing advanced the technology for dissipating energy in the Mars atmosphere and introduced the sky crane concept for depositing a heavy load (2 tons) on the surface of Mars. Further technical advances are needed for landing the masses of goal IV.

Data gathered by Curiosity will support protection of human biology from radiation at the lunar surface. The practicality of radioisotope thermal power for a long-term mission has been established. A great deal has been learned about the design of drive, suspension, and wheels for long distance driving on Mars (see Fig. 1.4).

Fig. 1.4 Artist's depiction of astronauts on Mars with spacesuits, backpacks for oxygen, a transporter, and equipment. (Image courtesy of NASA/JPL-Caltech)

Also, there is a related question of whether we can modify the environment of Mars to favor humanity. Curiosity can gather some data about the current and past environment, but experiments on modification are left to future missions.

Landing Site: Gale Crater NASA selected the MSL landing site to be on the floor of Gale crater, caused by a meteorite impact. The impact happened billions of years ago, near the edge of an enormous basin that covers nearly all the northern hemisphere of Mars. It has been proposed that this basin be called the Borealis Basin, the result of a very early impact event as Mars completed its aggregation as a planet. Meteorites that could cause such a large basin are sometimes thought to be more like planetoids or dwarf planets. The Borealis Basin may have once been flooded by an ocean in an early wet phase of Mars. If so, Gale crater would have been near its shore (see Fig. 1.5).

North of the Gale crater are several features that have the first name Aeolus, like Gale's central peak, Aeolus Mons (informally, Mount Sharp). In Greek mythology, Aeolus was the keeper of the winds. That's an appropriate name for features on windy Mars. The MSL Science Team selected the name Mount Sharp to honor a highly respected professor of geology, Robert Sharp, who had taught several members of the team. They were not aware that the International Astronomers' Union had already decided to name craters on Mars after deceased scientists and others related to Mars, but not to mountains in order to distinguish the classes of features. The MSL Science Team members have been free to

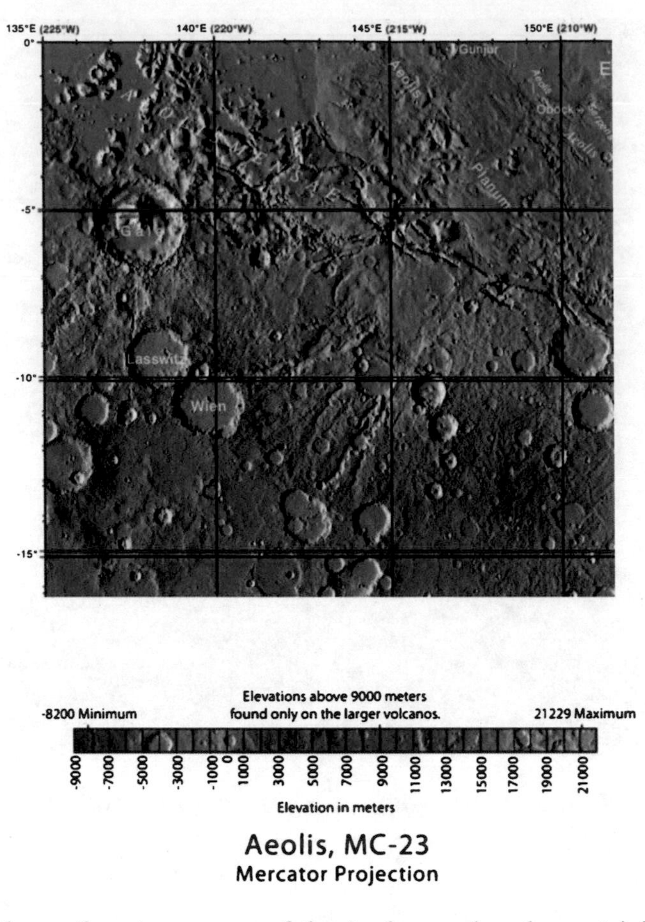

Fig. 1.5 Topographic map of the northwestern corner of the Aeolus quadrangle, containing Gale crater (upper left) and nearby Aeolis Mensae and Aeolis Planum. The highlands south of Gale are Terra Cimmeria. USGS made this nap for NASA from a Digital Elevation Map (DEM) based on the Mars Orbital Laser Altimeter (MOLA) instrument of NASA's Mars Global Surveyor (Image courtesy of NASA/GSFC/MOLA/USGS)

use either names in their peer reviewed papers. For more detail on this awkward issue, see an article by Kelly Beatty, "Mount Sharp or Aeolis Mons", Sky & Telescope, August 14, 2012.

South of Gale there is a steep rise in elevation that would mark the shore of the ocean in the Borealis Basin, if there was sufficient water. The steep rise is along the northern border of Terra Cimmeria, a very large highland region. Several channels on this steep slope are radial to the center of Gale crater and are thought to have brought liquid water to flood that crater through low points in its wall.

There is a very meaningful relation between Terra Cimmeria and Gale crater. In a period of heavy rainfall, water probably ran downslope from that very large highland and over the edge of Gale crater. In low spots of its edge, the water would have cascaded down to the crater floor, creating valleys. One of the younger of these valleys is called Peace Valley; its water was deposited as an alluvial fan (like a delta) on the crater floor. One stream of this fan flowed near the landing point of Curiosity! But that is getting a bit ahead of the tour. Let us proceed to a brief explanation of the geology of Gale crater (see Fig. 1.6).

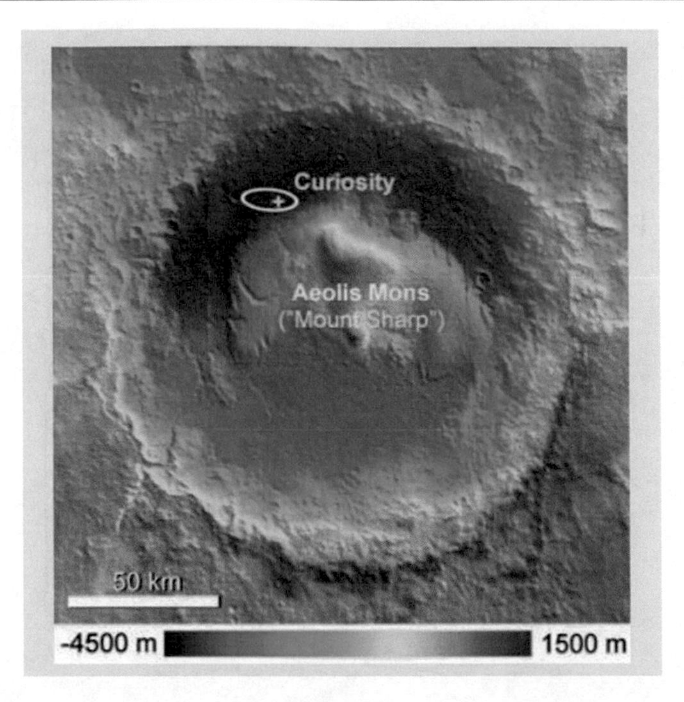

Fig. 1.6 The Curiosity landing target on the floor of Gale crater (Aeolus Palus) is shown in white, with the actual landing point in yellow. The latitude of Gale is 5.4 degrees South and the longitude is 137.8 degrees East. The diameter of the rim is 92 miles (156 km), and the peak of Mount Sharp is about 18,000 feet above the crater floor (Image courtesy of NASA/JPL-Caltech/GSFC/MOLA Science Team

Gale crater was formed by impact of a meteorite between 3.5 and 3.8 million years ago, judging by its degree of erosion and the relationship of its surrounding ejecta layer to nearby features. That age would correspond to the Noachian or earlier Hesperian global geologic ages. Based on orbital photography, including spectral data that distinguished minerals, Gale crater was formed of sedimentary rock, with windblown sand covering part of its surface. The southern half of the rim appears higher, relative to the floor, than the northern rim, which could be due to water covering Borealis Basin at the time of impact. It could also be due to its impacting meteoroid approaching a little north of vertical.

The central mound within Gale (Aeolis Mons, officially, but Mount Sharp, informally) appears to be covered with a succession of sedimentary layers of different minerals. Central peaks are not unusual for impact craters of this size, but this one seems to have acquired heavy deposits of sediment from external flooding that have been eroded to leave an irregular mound (Fig. 1.7).

Figure 1.7 combines data from three orbiters, elevation from ESA's Mars Express, high-resolution images from NASA's Mars Reconnaissance Orbiter, and color from NASA's Viking Orbiter.

Gale was chosen as a landing site for Curiosity because of several factors indicating the likely presence there of ancient liquid water. Concentric channels from the highlands to the south crossing low points on the rim were seen from orbit to lead to valleys on the floor and fans of material. Of course, these features are dry now, but they could have delivered a great deal of water in the past, draining from the highlands. Among the bands of diverse minerals on Aeolis Mons are clay and sulfates. Clay has been suggested as a medium for the origin of life. Sulfates could contribute atomic sulfur, known for preserving organic carbohydrates of past life.

These qualities have made a delightful playground for Curiosity's exploration, as we will see as we tour its route.

Fig. 1.7 Vertical view of Gale crater, showing the central peak surrounding a complex mound of sediment, formally named Aeolus Mons but widely known as Mount Sharp. (Image courtesy of NASA/JPL-Caltech/DLR/FU-Berlin/MS. Image obtained from JPL Photojournal, PIA 15687)

Mission Summary

The MSL mission is a challenging one. The size and weight of the Curiosity rover has been increased to accommodate a large array of instruments. These require a large amount of power. That power is supplied by a thermoelectric generator to assure long-time operation that takes into account the capricious Martian winds and dust. The diverse set of instrumentation is worthy of a stationary laboratory facility on Earth. The increase in weight has forced a more complex design of the sequence of descent and landing that is potentially scalable for the demands of human exploration. These increases in capability have resulted in a profound new understanding of Mars, even while the Curiosity rover is still climbing Mount Sharp, entering new geologic terrains as scores of scientists analyze the data to date and publish their results. The adventure continues! And it will continue with the next Mars mission.

2 Curiosity's Resources

Curiosity is the most recent addition to a series of Mars rovers. Mars Pathfinder's rover Sojourner was first, followed by the twin Mars Exploration Rovers Spirit, and Opportunity (see Fig. 1 from Preface). At least one of these has been operational since 2004. As the size and weights have increased, so have the number of instruments and the sophistication of the technology, supported by growth in operational experience and knowledge of the Mars environment.

The record for longevity and distance traveled is held by the Opportunity Rover that was still exploring the Endeavor crater when Curiosity landed. Opportunity traveled 45.16 km. in 14 years and 46 days. Will Curiosity set a new record? It is not obvious, because Opportunity traveled much of its distance going to the Endeavor crater without many distractions on the way, traveling through fairly uniform terrain. Powered by solar panels periodically dusted by winds, its mission was terminated by a global dust storm that not hurt Curiosity's Radioisotope Thermoelectric Generator (RTG).

Curiosity benefits not least from the experience of the designers and operations teams. It is not easy to survive in the Mars work environment, including the two-year launch cycle and the periodic asynchrony between Earth days and Mars sols. It took time for the operations teams to adjust to the rhythm of planning, reviewing data, and modifying the plans.

The increase in payload weight of Curiosity relative to all previous rovers allowed for a striking increase in the number and sophistication of the science investigators and the instruments they designed, tested, and operated. They came from a wide variety of American and international government agencies, universities and private firms, and bring experience in space operations. These scientists, about 500 in number, also analyze the voluminous data Curiosity generates and publishes peer-reviewed papers as the mission continues. This author attended a large gathering for the landing of Curiosity. The chant started up, "USA, USA …" and the crowd spontaneously shifted to the chant "JPL, JPL …" in recognition of the truly international nature of the achievement.

JPL was started by Caltech under the sponsorship of the U. S. Army during World War II and transitioned to NASA. The Principal Investigators for Curiosity are led by current and past professors from Caltech, and Mount Sharp is (unofficially) named after Robert Sharp, a former professor of geology at Caltech. The International Astronomical Union has formally named the Robert Sharp crater, located not far from the much younger Mount Sharp.

C. J. Byrne, *Travels with Curiosity*, https://doi.org/10.1007/978-3-030-53805-7_2

Curiosity has the benefit of there being several Mars orbiters that can be used for two very important support contributions. First, there is the systematic and coordinated imagery and comprehensive mapping (including topographic and geologic maps) that are produced from orbital observations. These products are used for route planning and target selection. Second, the communication relay function performed by the Mars orbiters permits the floods of imagery data from the multitude of cameras on Curiosity, along with the data from the other scientific instruments, to be transmitted at a tremendous savings of weight and power for Curiosity.

The flood of data depends on the continued growth of the contribution of NASA's Deep Space Network, which was designed, built, and operated by JPL at the outset of Moon and planetary exploration by spacecraft in the early 1960's. Since the Earth is turning, a location anywhere else in the solar system is in view from a specific location on the surface of Earth for only a few hours. Continuous surveillance of a spacecraft requires communication systems at three locations around Earth. JPL established those three locations very early on, near Barstow California, near Canberra, Australia, and near Madrid, Spain. These three locations are roughly 120 degrees of longitude apart, so as a spacecraft sets towards the horizon at one, it has risen at another. It is a difficult scheduling problem for the operations team to operate the big antennas to cover all the spacecraft touring the solar system, some closer to the sun than Mercury and some further from the sun than the heliopause.

Elements of personality have been attributed to Curiosity, such as a practice of using feminine pronouns in reference to this complex robot with a flexible program. Artificial intelligence concepts are used for hazard avoidance and goal seeking. Yet when a group makes rule changes by consensus, the resulting behavior by Curiosity in response to its environment can produce a feeling of "personality."

Raw data is published by JPL as it is received, often on a daily basis. Calibrated and analyzed data is published on an approximately yearly basis, depending on the timing of peer reviews. In between, abstracts are published, often by several co-authors. New ideas have frequently arisen, especially as Curiosity has climbed into new geologically different areas.

As alternate hypotheses have been generated, groups of scientists with diverse experiences have proposed new theories. This process has involved new plans for Curiosity to change routes and observations to resolve the differences of opinion and reach a consensus. Examples of this process will be mentioned in the course of following the journey of Curiosity, emphasizing the interplay of observation and idea generation made possible by the extended duration of the mission.

A special contribution of the scientific community is the invention and development of new instruments and sensors for Curiosity, including testing, calibration, and operations throughout the mission. The special subject of Principal Investigators will be discussed in a further section.

Like Spirit and Opportunity before it, Curiosity has six wheels (see Fig. 1 from Preface and Fig. 2.1 and Fig. 2.2), with individual electric motors and flexible suspensions and axles to enable driving over very rough rock. The wheels and suspension have been scaled up to support the expanded scientific payload and also the heavy Radioisotope Thermoelectric Generator (RTG), which stores enough radioactive material for up to 10 years of power for mobility, heating, and instruments.

Fig. 2.1 Mobility test of wheels and suspension off-weighted to simulate mars gravity. The location is in the dunes of Death Valley. (Image courtesy of NASA/JPL-Caltech)

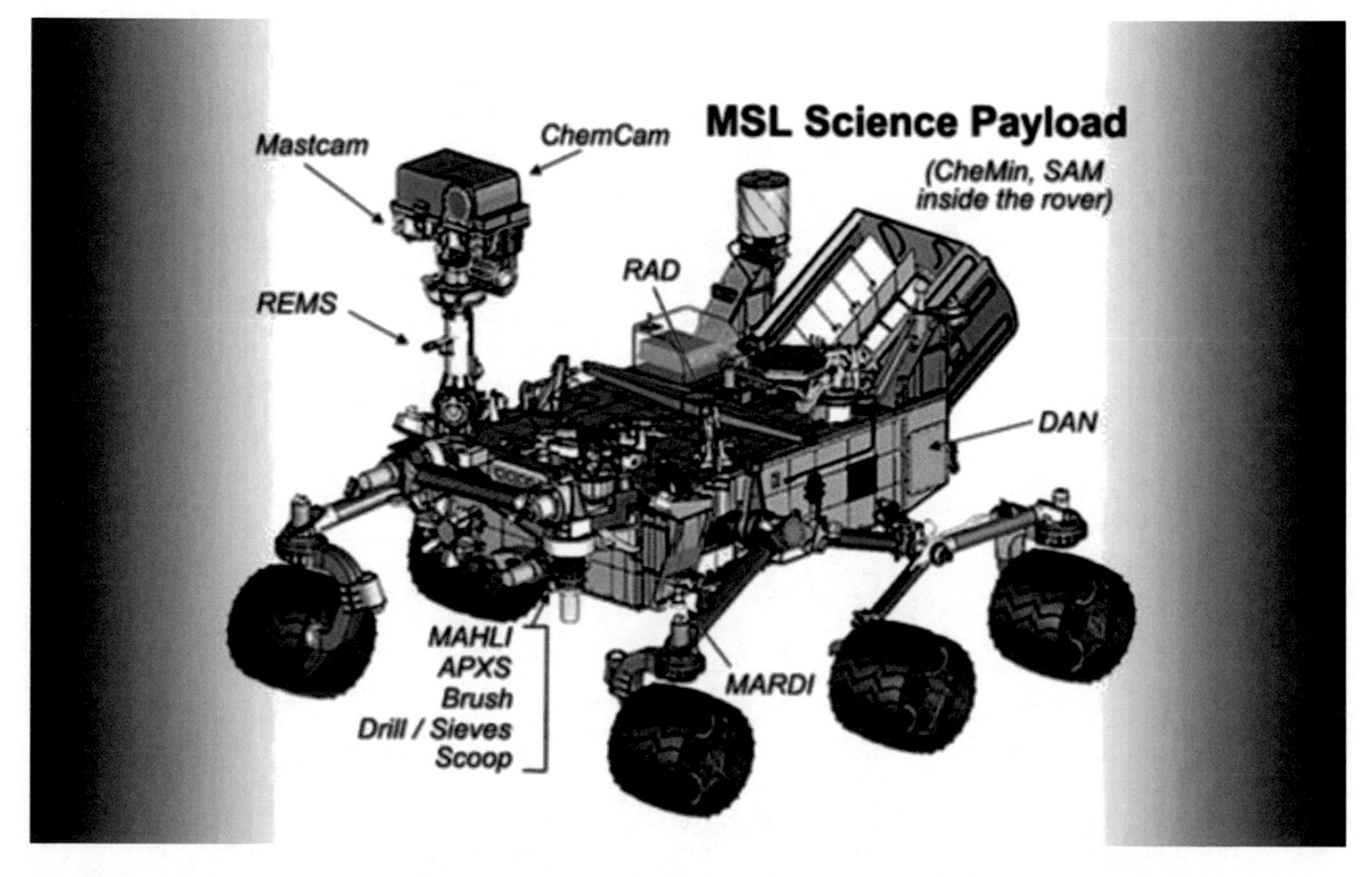

Fig. 2.2 The science instruments carried by the curiosity rover. (Image courtesy of NASA/JPL-Caltech)

Mobility starts with the wheels, power and suspension, and JPL continues to work on problems encountered on Mars, testing new ideas to counter problems with experiments in different Earth environments and simulation chambers. Getting stuck in sand is one problem. Lubrication is another. Sharp stones are another. Wheel slip is yet another.

In off-road vehicles, keeping the "unsprung weight" low is important. After encountering an obstacle, a rotating tire rises and is pushed down by the suspension. The less momentum in the wheel, the smoother the ride.

Mobility alone is not enough to assure effective exploration; you also must be able to see where you are. Two cameras are mounted low in the front and in the back. They provide stereo views for distance to support a hazard avoidance function and are named the Left and Right Hazcams. They also provide images that support deployment of instruments mounted on the arm for surface observation and drilling for samples.

Four additional cameras, the Left and Right pairs of Navcams, are mounted higher in the front, with longer focal length for navigation, seeking targets of interest and lower-slope paths to reach a goal. They are often used for wide-angle mosaics to survey the surroundings, a function critical to "being aware of the situation." The Navcams are provided in pairs for each side for redundancy. One of each pair is cabled to the A computer and the other to the B computer, so that if one of the Navcams fail, stereo vision can be restored by switching to the active computer. If a computer fails but all Navcams are operational, the working computer will still have stereo vision.

All these cameras (17!) produce a massive amount of data that is stored in flash drives until an orbiting satellite comes into view. Then the data is sent there for transmittal to Earth. This is an example of the close support provided to Curiosity by orbital spacecraft who also receive targets for high resolution images and spectroscopic geology data for maps of the exploration area available to Curiosity's planning teams. Such information has been very valuable for the selection of the Gale Crater and also for planning safe routes to goals of scientific interest.

The scientists who have proposed instruments and sensors are one of the greatest strengths of the MSL operations. Having designed and tested the instruments before assembly, they activate and calibrate them after landing. During operations for Curiosity, they select their science targets and participate in the day-to-day planning meetings. The scientists review the raw data returned each day, analyze it, publish peer-reviewed papers and abstracts regularly and participate in meetings in their disciplines. The MSL Project Science Group (PSG) advises the project on optimization of mission science return and resolution of science issues. During landed operations, the PSG provides strategic advice to the Science Operations Working Group. The original list of PSG members is shown in Table 2.1. In January, 2015, John Grotzinger stepped down as project scientist to become chair of the Divisiion of Geological and Planetary Sciences at Caltech. Ashwin Vasavada (JPL) became the new MSL Project Scientist.

Table 2.1 MSL project science group members, including principal investigators (PI)

Project science group role	Name	Institution
Project scientist	John Grotzinger	Caltech
Program scientist	Michael Meyer	NASA Headquarters
APXS PI	Ralf Gellert	University of Guelf, Canada
ChemCam PI	Roger C. Wiens	Los Alamos National Laboratory
CheMin PI	David F. Blake	NASA Ames Research Center
MAHLI PI	Kenneth S. Edgett	Malin Space Science Systems
Mastcam and MARDI PI	Michael C. Malin	Malin Space Science Systems
RAD PI	Donald Hassler	Southwest Research Institute
REMS PI	Javier Gomez-Elvira	Centro de Astrobiologia, Spain
SAM PI	Paul Mahaffy	Goddard Space Flight Center

Table 2.2 Science instruments and acronyms

Group	Acronym	Instrument
Cameras	Mast camera	MastCam
	Mars hand lens imager	MAHLI
	Mars descent imager	MARDI
Spectrometers	Alpha particle x-ray spectrometer	APXS
	Chemistry and Camera	ChemCam
	Sample analysis at Mars instrument suite	SAM
Radiation detectors	Radiation assessment detector	RAD
	Dynamic Albedo of neutrons	DAN
Environmental sensors	Rover environmental monitoring station	REMS
Atmospheric sensor	MSL entry, descent, and landing instrument	MEDLI

Science instruments are the payload for the Mars Science Laboratory; the motivation for the mission. Some instruments make near or actual contact with the surface rocks or sand. Others are remote sensors. When Curiosity enters a new area, the scientists typically ask for a "walkafbout" using remote instruments to get a general impression, and then select locations for contact examination. An additional group of scientific instruments are the environmental sensors, which operate nearly independently of the other instruments except for special studies (Table 2.2).

The instruments work together with the engineering equipment. For example, a core sample is taken from a stable rock with the drill tool mounted on the turret at the end of the arm. The drill works by either rotation, percussion, or both together. It can drill to 6.5 cm depth. It takes a core sample about 1.6 cm in diameter. Portions of the sample are sieved and distributed to the Sample Analysis at Mars (SAM) suite of instruments and the Chemistry and Mineral X-ray Diffraction and Florescence (CheMin) instrument. The sample handling system is called Collection and Handling for In Situ Martian Rock Analysis (CHIMRA).

CheMin, as its full name implies, produces abundances of crystalline grains of the sample by element and crystal structure (see Fig. 2.3). An X-ray beam is directed at a portion of the drilled sample. The

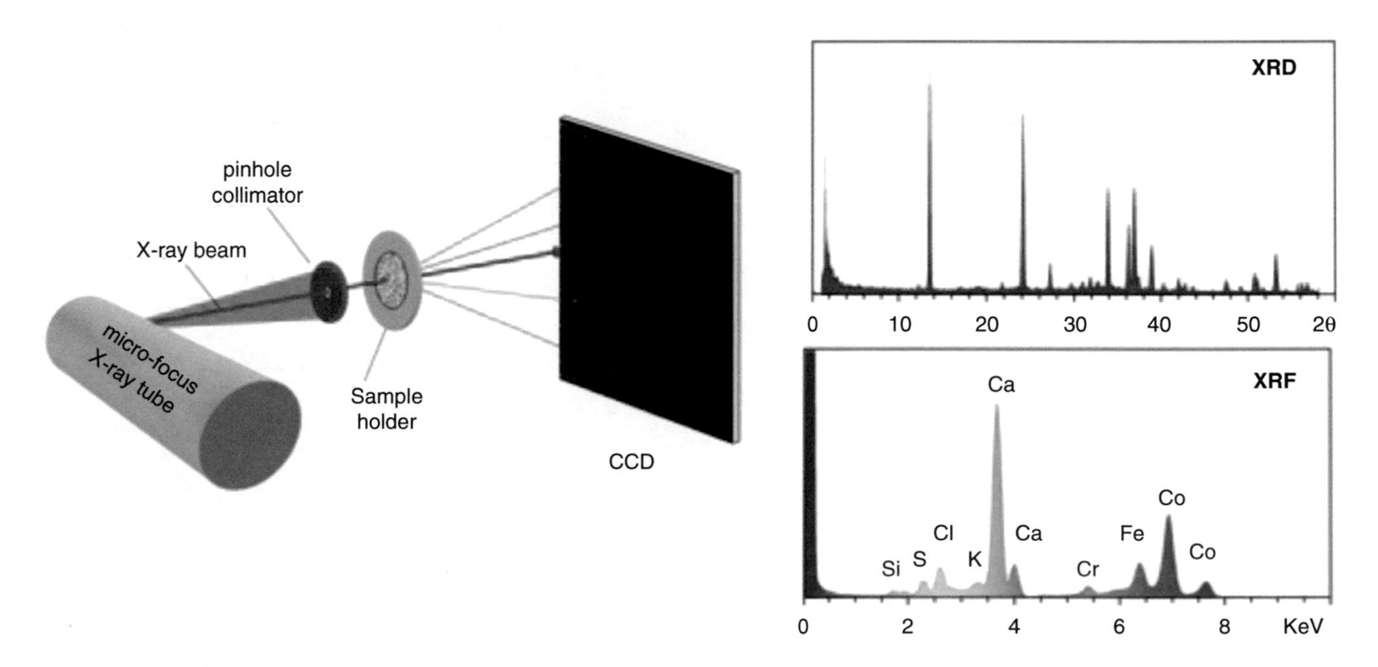

Fig. 2.3 **CheMin schematic.** An X-ray beam is collimated by a pinhole and illuminates a sample. The resulting diffraction pattern is analyzed to form a distinctive pattern for major minerals (XRD graph) and a spectrogram showing the elements (XRF), including for the mudstone (amorphous) part of the sample. (Image courtesy of NASA/JPL-Caltech, Science Corner)

Fig. 2.4 The first diffraction pattern by CheMin taken from a sample of Mars soil scooped from the patch of sand at the Rocknest location. This was the first of Curiosity's scoop samples. (Image courtesy of NASA/JPL-Caltech)

pattern of diffracted X-rays and florescence is analyzed to reveal the elements of the structure of crystalline grains. Amorphous material (lacking crystal structure) is termed mudstone by geologists; such material is an indicator of sedimentary rock deposited by water.

A raw diffraction pattern from CheMin is shown in Fig. 2.4.

The image in Fig. 2.5 of CheMin illustrates how rugged an instrument's structure must be to retain the precision accurate measurements require.

Fig. 2.5 This image shows the design model of the CheMin instrument upside down on a work bench. The alignment bench is on the top when the instrument is deployed. It is stiff enough to keep the X-ray tube, sample wheel, and CCD assembly aligned within a few tens of microns during a sample analysis

SAM, the most complex suite of instruments (see Fig. 2.6) first produces a beam of vaporized molecules, hits them with an electric charge, and then deflects them with a designed magnetic field. The resultant displacement of the molecules is measured to infer their precise molecular weights. In combination with CheMin results, the environmental conditions under which the rocks were formed and modified (such as atmosphere composition and pressure and the Ph of water) can be estimated.

SAM is also used for periodic analysis of air samples. Its precision concerning molecular weights is sufficient to infer atomic weights of light, simple molecules. Of particular interest is data about isotopes of oxygen because they offer clues to Mars' origin (Goal III of the Mars Exploration Program).

APXS is a contact instrument on the turret. Placed within 2 cm of a rock, APXS irradiates it with alpha particles and X-rays, inducing the emission of X-rays and florescence. A spectrogram of the emissions reveals the elements presence in the sample and the peaks and valleys of the spectrum show the relative abundance of each element.

Figure 2.7 shows the three parts of the APXS instrument, the sensor head, the electronic unit, and the calibration target, each mounted separately on Curiosity. The data from the APXS was valuable for comparative analysis of Murray Formation rocks when solid samples could not be obtained while the drill was inoperable during part of the ascent.

Fig. 2.6 Photo at Goddard Space Flight Center of the SAM suite of instrument for the analysis of both solid and atmospheric gas samples. SAM was being assembled, when his photo was taken, with its side panel not yet in place. (Photo and labels courtesy of Goddard Space Flight Center. Image is PIA 16100, courtesy of NASA/JPL-Caltech.)

Fig. 2.7 The sensor head (right) irradiates the sample with X-rays and alpha particles and its 2-dimenstional sensors read the resultant APXS backscatter and fluorescence. The signals are received by the electronics unit (left) and sent to earth for display. The calibration target (center) is of known elements and minerals used to compensate as sources and distances change

Remote Instruments

The following instruments work at a distance from the rock and cover a wider area:

ChemCam is an active remote instrument that fires a high-powered short burst from a laser at a rock surface, producing a flash of hot plasma. A telescope catches photons from the flash, and the spectrum of the flash indicates the presence and relative abundance of the elements. The resulting table can be compared with APXS (Figs. 2.8 and 2.9).

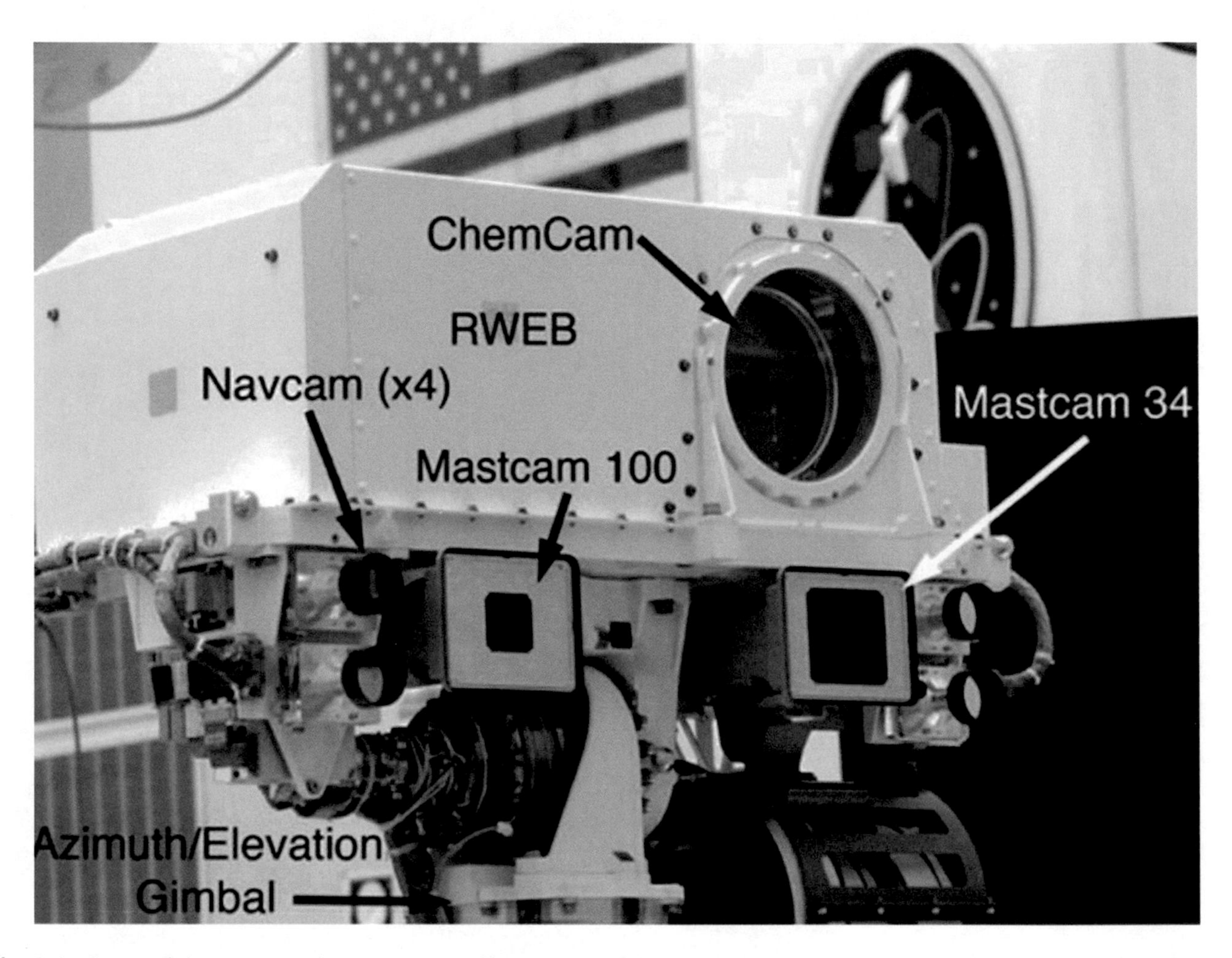

Fig. 2.8 Some of the remote science cameras, Chemcam, and the two Mastcams, are mounted high on Curiosity's Remote Sensing Mast so that they can see over ridges and low hills. The Remote Warm Electronics Box (RWEB) is about 9 feet above the surface. It provides a line of sight well above that of a walking geologist. The mast also carries the four Navcams. Although these are engineering cameras, essential for navigation, their images are also useful for the scientists, who can request images from the Navcams. The Azimuth/Elevation Gimbal has a range of 360 degrees in azimuth and nearly 180 degrees in elevation

Fig. 2.9 Flight model ChemCam instrument. The mast-unit (*top*) is comprised of the laser, imager, telescope, and focus laser. The body-unit (*bottom*) includes an optical demultiplexer, three spectrometers, the central processing unit and a thermo-electric cooler

Mastcam is comprised of two full-color cameras that are mounted on the Remote Sensing Mast about 1.9 m (6 feet) above the ground surface for visibility above low ridges. One camera is like a typical consumer 35 mm camera and the other is like a telephoto camera (see Fig. 2.10).

Fig. 2.10 The two Mastcams are nearly identical except for the optics. Left: 34 mm focal length with a 15° field of view. Right, 100 mm focal length with a 5.1° field of view. The 34 mm Mastcam can be focused from 0.34 mm to infinity and the 100 mm Mastcam from 1.63 mm to infinity. They each deliver 1280 by 720 pixel images, with high definition video at eight frames per second. The jackknife is there for scale (88.9 mm long)

The primary objectives of the Mastcam cameras are to characterize and determine details of the history and processes recorded by the geology the Gale crater field site. This includes but is not limited to observations of landscape that enable understanding of past and present geologic processes; studies of frost, ice, documentation of clouds, dust devils, storms, and other atmospheric events; and assist contact instrument science. They can be used individually or for stereo images; the stereo base is 24.5 cm.

Curiosity, being on Mars, has to contend with the iron-rich dust in the atmosphere, which tints the sunlight to a dull red, and that tints the entire scene. To humans, this is a source of confusion. So sometimes the images are spectrum-shifted (after being received on Earth) to simulate clearer atmosphere that we are used to seeing.

MAHLI, the hand lens imager, has as its primary responsibility examining rock samples to measure grain size (see Fig. 2.11). It is a color camera, attached to the turret at the end of the arm, and therefore can be aimed where no other camera can reach, such as above the mast and below the rover. Part of the family with the two Mastcams and MARDI, it has autofocus ability and is unlimited in how far it can see. Consequently, it is a jack of all trades, limited mainly by the obvious restriction that while any other instrument on the turret is active, it cannot be used. It is really a transition instrument, with both remote and contact capability. See Fig. 2.12 for its physical structure.

The MAHLI instrument has three parts, the focusable color camera head, a digital electronics assembly, and a calibration target. The camera and the target are on the turret and the centralized electronics are in the rover body. The camera head is shown in Fig. 2.9.

Fig. 2.11 Test image of a zinc ore sample taken by the MAHLI (hand lens). Note the color quality, glints off the crystal faces, and the 1 mm scale bar at lower right corner. (Image courtesy of NASA/JPL-Caltech, Science Corner)

MARDI, mounted within Curiosity looking downward, provided very valuable engineering data during the powered descent and the sky crane phases of EDL (see Chap. 3). MARDI took a continuous movie at four frames per second from parachute deployment, through heat shield release, parachute release, powered descent, and sky crane operation, to about 2 minutes after landing. The movies showed the behavior of MSL as it passed through the atmosphere and reacted to changes in density and the ambient wind. The sequence of frames showed where Curiosity landed within the target area, critical data for determining the strategic path Curiosity would take for the next year. (see Fig. 2.13 for the physical appearance of MARDI).

After landing, MARDI took pictures of the surface directly below Curiosity's left side. As the mission progressed, software changes were made to provide "sidewalk mosaics" of Curiosity's path, showing continuous patterns of the Murray Formation's fractured rock plates. Usually, the pictures were taken in twilight, to avoid high contrast between sunbeams and shadow.

Fig. 2.12 MAHLI replaces the hand lens of the traditional geologist, and if the turret on the arm is an analog of a hand, is at least as flexible in examining a rock or determining whether your equipment is dirty. For example, it is useful for taking a rover self-portrait or checking the wheels for damage. The camera head unit contains an optomechanical assembly, a focal plane assembly and the camera head electronics assembly. The pocket knife, 88.9 mm long, was included for scale and maybe to invoke MAHLI's versatility. It has white LEDs for night photography and UV LEDs to excite fluorescence

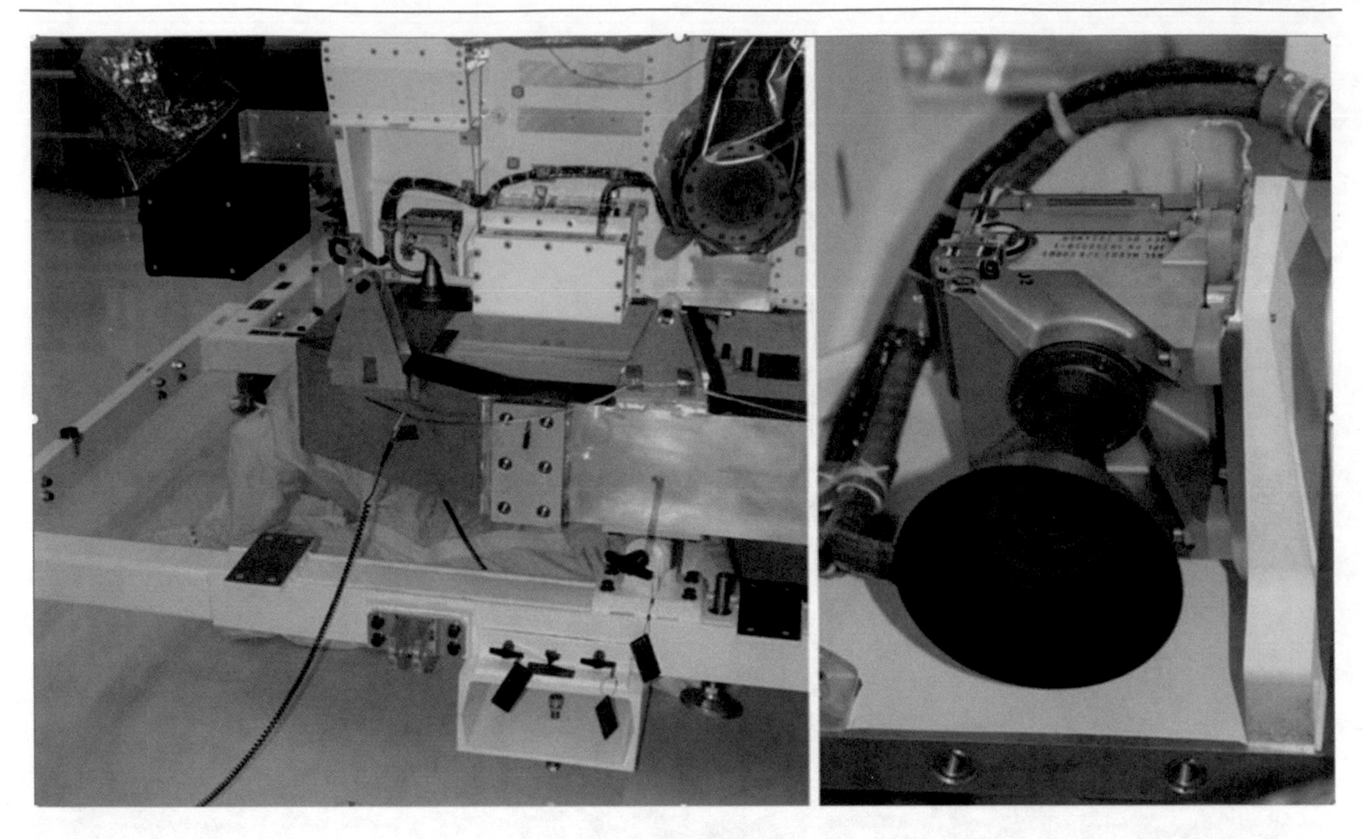

Fig. 2.13 MARDI is a member of a close family, with the two Mastcams and MAHLI cameras having the same electronics and platen units. The optics and the mounts to the rover are what makes differences in specifications. Only MARDI has a fixed mount: it always looks down, relative to the rover frame

Environmental Instruments

The following instruments contribute to Goals I, III, and IV of the Mars Exploration Program.

RAD measures the spectrum of energetic particle radiation from the surface of Mars. This includes solar energetic particles, galactic cosmic rays, secondary neutrons, and other particles generated in the atmosphere or ground of Mars (see Fig. 2.14).

Because Mars lacks a strong magnetic field, like Earth has, ambient radiation is a greater concern to organisms, including humans. The RAD data is very important for human stay time on Mars, depending on the shielding provided (Goal IV).

The physical RAD unit is shown in Fig. 2.15. It is mounted on the exterior of the rover.

REMS is designed to measure wind speed and direction, pressure, relative humidity, air temperature, ground temperature, and ultraviolet radiation. Measurements are taken several times per day. Curiosity adds surface weather data to global models derived from observations from orbital spacecraft (Goal III).

Loss of the wind speed and direction sensors during the landing was unfortunate because wind is the driver of sand and dust and an important parameter in dune dynamics. Some wind information was obtained by comparing images to measure the movement of sand. The two REMS mounting beams are shown in Fig. 2.16.

DAN fires pulses of neutrons into the surface below Curiosity. The neutrons rebound as they hit hydrogen and OH ions within 0.5 m of the surface and are detected by DAN, which estimates water bound in rock by hydration of minerals. The neutron pulse generator and sensor are shown in Fig. 2.17. This instrument operates at frequent intervals as the rover is traveling to its next goal and makes a record of detected hydrogen as it goes. It reports the data as equivalent water molecules (Goal I). The DAN Instrument was especially valuable during the Marias Pass walk-about when it picked up a hot spot of

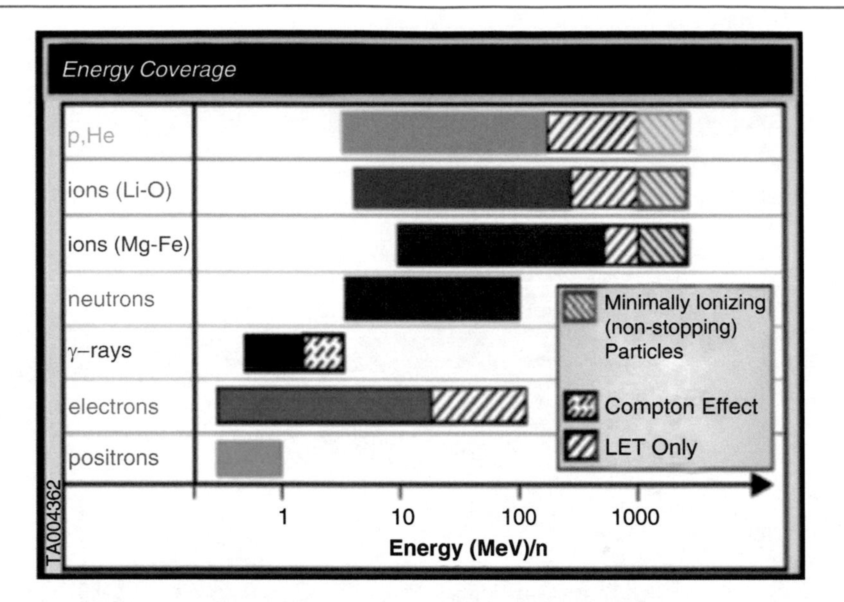

Fig. 2.14 RAD energy coverage for both charged and neutral particles. (Chart courtesy of NASA/JPL-Caltech)

Fig. 2.15 RAD's self-contained unit in the lab. It's charged particle tunnel has a 65° field of view pointing towards the zenith (when Curiosity is level relative to gravity). Orbital spacecraft can measure the solar wind directly, but Rad measures that part of the solar wind that penetrates the thin Mars atmosphere, plus secondary radiation

hydrogen, probably bound in the subsurface minerals. In combination with evidence from other instruments indicating a high level of silica in fractures in the Murray Formation rocks, a clear story emerged of diagenesis from the overlying Stimson Formation.

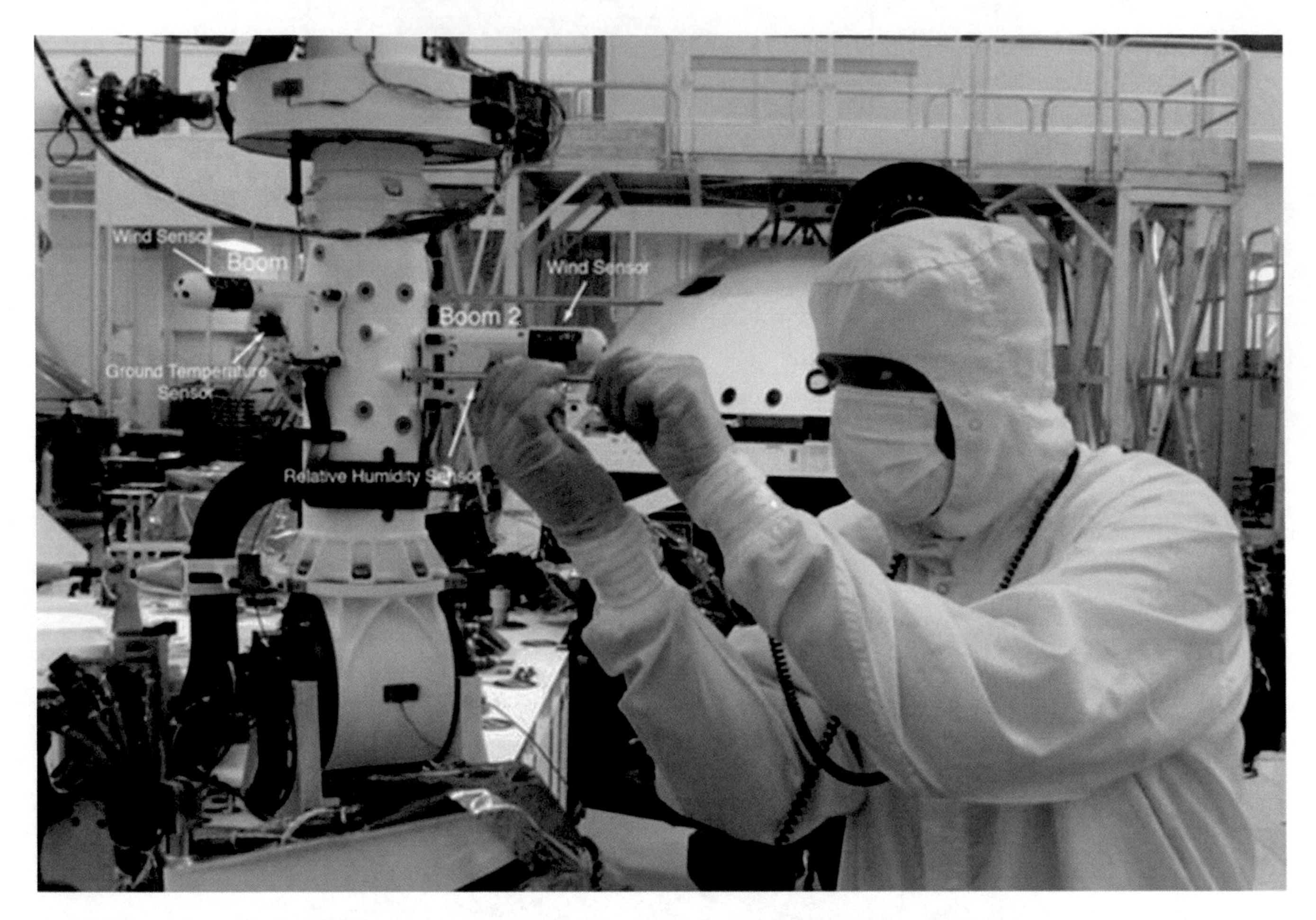

Fig. 2.16 REMS flight booms are being installed on a section of Curiosity's mast. The engineer's hands are on Boom 1, which holds wind and ground temperature sensors and across the mast is Boom 2 with wind and humidity sensors

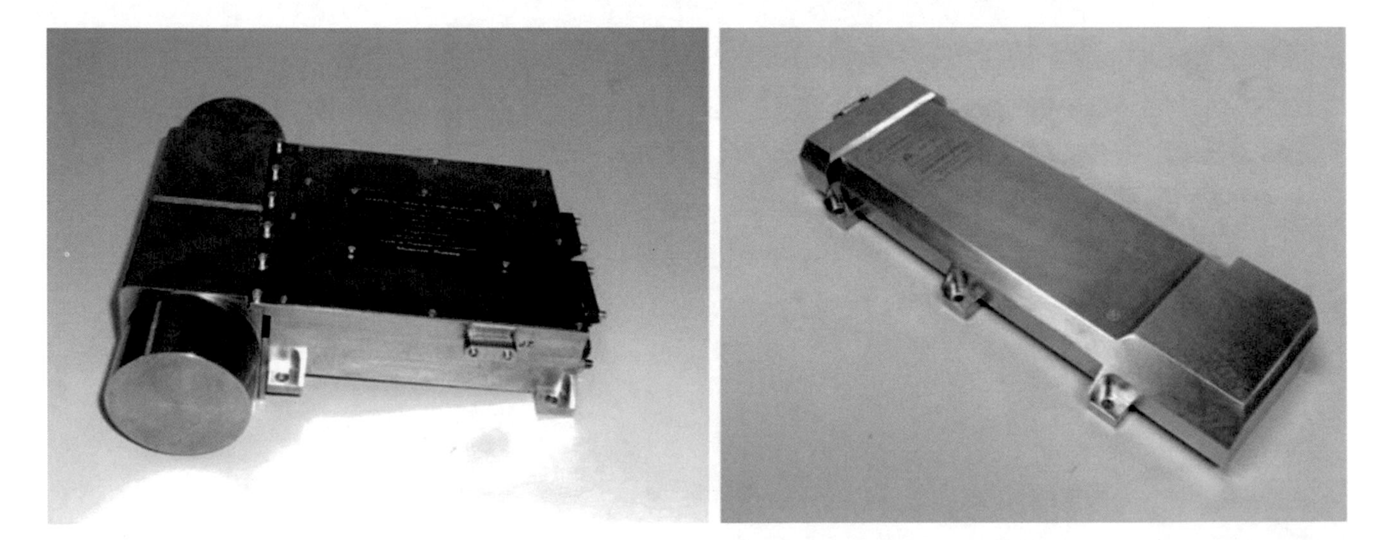

Fig. 2.17 DAN's detection sensor is on the left and the pulse neutron generator is on the right. Some of the neutrons encounter nuclei of atoms in the surface and may bounce back to the sensor, which measures their energy. Such neutrons that have lost the least energy have bounced off a single proton, the nucleus of a hydrogen atom. Since the neutron beam is pulsed, the time of arrival of a neutron to the sensor can be used to infer the depth of the hydrogen atom

3 Entry, Descent, and Landing

Previous landings on Mars had been by ablative shield and parachute followed by rockets or air bags. It was decided to change this for the MSL, allowing a much heavier 2 ton weight of the payload and the desire for even heavier payloads in the future (see Fig. 3.1).

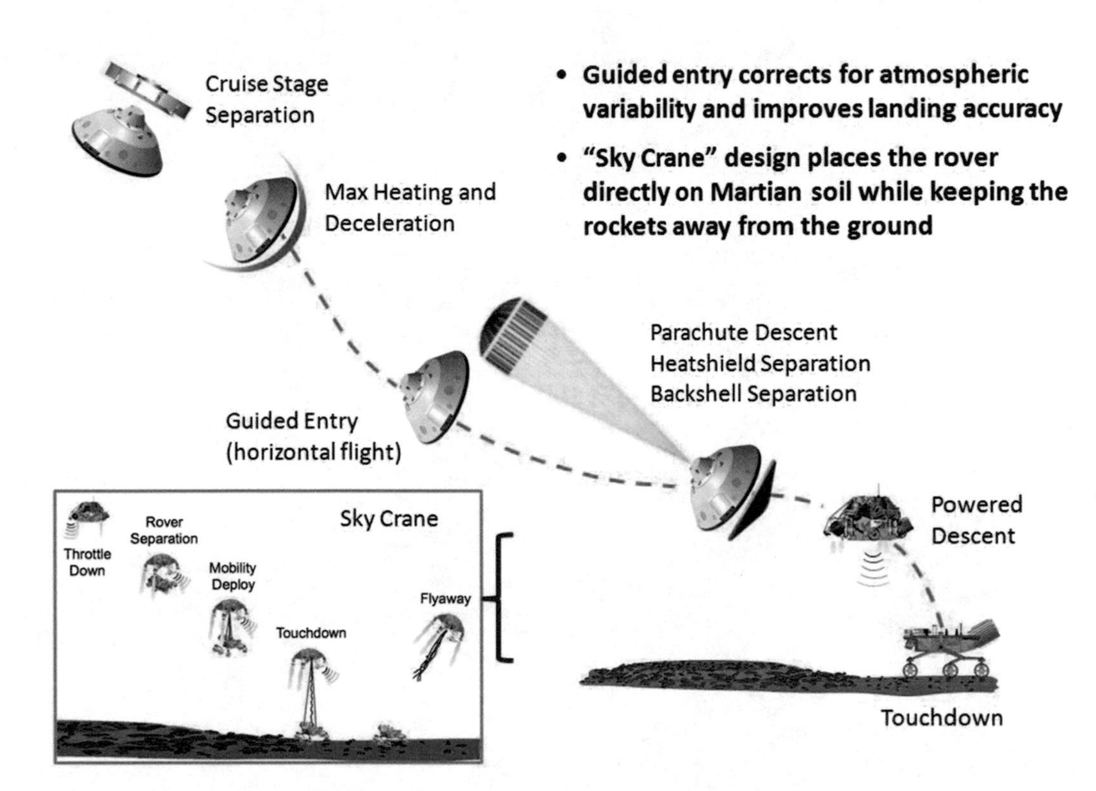

Fig. 3.1 Diagram of the sequence of operations during the descent and landing period of the MSL, showing the deployment of systems used to dissipate the energy left after the entry period (see text). The thin atmosphere of Mars called for a series of additional steps to bring the payload to a soft landing. After parachute deployment, there would be a series of jettisons of equipment no longer needed to reduce total momentum. This continued the actions of dropping mass that was begun before entry with the cruise stage. A rocket-propelled descent stage would drop the elevation and slow the speed to a soft landing. (Image courtesy of NASA/JPL-Caltech)

C. J. Byrne, *Travels with Curiosity*, https://doi.org/10.1007/978-3-030-53805-7_3

This time there was to be a sky crane to aid in a rocket-propelled descent. After the rockets slowed the descent stage to hover, sky crane would lower Curiosity to the ground on a cable. As Curiosity sensed that the tension on the cables dropped, it would realize it had landed and cut the cables and a communication line. The descent stage would fly off and crash to the surface a safe distance away.

The first step in the actual descent and landing was to disable the fault management software to avoid a temporary overload or other transient event from switching processors during a critical part of the mission. The initial entry velocity was 5,800 m/s. The active aeroshell and its heat shield (Fig. 3.2) dropped as much of this energy as possible, considering the thin atmosphere of Mars. Then, as the Mars atmosphere became less tenuous, the ablative shield absorbed an increasing amount of energy. In this phase, a shifting mass was used to adjust the attitude of the spacecraft to keep the heat shield aligned with the relative velocity of the spacecraft within the atmosphere.

Fig. 3.2 The MSL assembled for testing. The Aeroshell with its dark heat shield is below the rectangular radiators on the circular cruise stage. They are parts of the equipment that is jettisoned before entry into the Mars atmosphere. Curiosity is inside the Aeroshell. (Image courtesy of NASA/JPL-Caltech)

As the initial energy dropped, the atmospheric cruise phase began. The attitude was adjusted to use the spacecraft, with its shield, as a lifting body. In this phase, under inertial guidance, the spacecraft followed a complex course to a designated target in latitude, longitude, and elevation while also losing energy, a strategy also followed by the Viking spacecraft and the Space Shuttle.

At this point, the speed dropped to about 400 m/s and the density of the atmosphere allowed a very large parachute to be deployed (see Fig. 3.3).

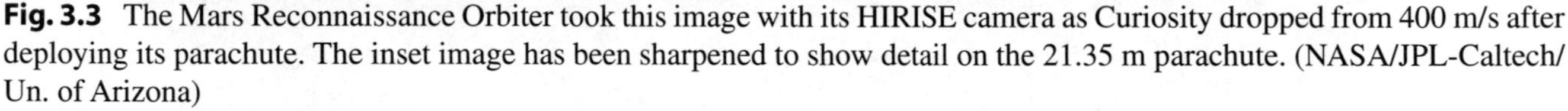

Fig. 3.3 The Mars Reconnaissance Orbiter took this image with its HIRISE camera as Curiosity dropped from 400 m/s after deploying its parachute. The inset image has been sharpened to show detail on the 21.35 m parachute. (NASA/JPL-Caltech/Un. of Arizona)

In this phase, equipment no longer needed had been dropped, including the heat shield and attitude masses. Radar is now in contact with the surface.

Now, warm up descent rocket engines. Release parachute and backshell. Ignite descent rocket engines to absorb the rest of the entry momentum. During this stage, wind would contribute to the drift from the center of the target area.

Under the power of rocket engines, and with radar guidance for elevation, attitude, and ground speed, the decent stage lowered to zero horizontal motion and a low rate of descent near the surface.

The sky crane on the descent stage (Fig. 3.4) lowered the payload to the ground on three cables and an umbilical cable (for communication). The descent stage dropped at a nominal 0.75 m/s as Curiosity's weight pulled out the cables (braked in the descent stage) until they reached their full length of 7.5 m. Meanwhile the suspension and wheels were unfolded, dropped by gravity, and locked in place. If, like Curiosity, a future landed payload is a ready-to-roll rover, it will land gently on its compressible wheels; otherwise, there will be some other system to soften the touchdown, like compressing legs or shock absorbers.

When the payload sensed that the cables had gone slack, it concluded it had landed and cut them and the communications line. The descent stage with the sky crane flew a safe distance away when the cables went slack and crashed.

After the challenging problems in development and testing this mode of moving from space to the surface of Mars, the combination of technologies used to dissipate the energy of space flight worked as designed, establishing the feasibility of this scalable mode of soft-landing a fragile payload. Bradbury Landing, on August 6, 2012 was about 2.3 km from the center of the target area, without a beacon to guide it or hazard avoidance. The landing site and the geological formation of the surrounding crater floor were named after Ray Bradbury, a science fiction writer who passed away 2 months before the landing. Ray Bradbury wrote the novella "Martian Chronicles" about refugees that left Earth, after it had suffered a nuclear war, to colonize Mars. At the time, 1950, the cold war was beginning, and there was widespread fear that gave the topic credibility. Many members of the MSL Science team had become interested at a young age in space technology through science fiction.

Fig. 3.4 The descent stage becomes the sky crane, lowering Curiosity, with its wheels deployed and computer powered up, slowly to the ground as in this image from a test. When Curiosity senses touchdown, it cuts the cables and umbilical cable. When the descent stage senses no tension in the cables, it throttles up four of its rockets (not directed at Curiosity), providing enough thrust to send itself several kilometer away to avoid contact with Curiosity. (Image courtesy of NASA/JPL-Caltech)

As MSL delivered Curiosity to the surface, the Mars Odyssey Orbiter was in the proper place in its orbit to receive telemetry from the MSL and relay it to Earth throughout the 7 minutes from entry to touchdown, and to relay the early telemetry from Curiosity.

Mars Reconnaissance Orbiter, that had taken the parachute photo in Fig. 3.3, took a picture of Bradbury landing (see Fig. 3.5), showing the various elements of the MSL that had brought the Curiosity rover to a safe landing, ready to flex its wheels, commission its instruments, and roll.

Fig. 3.5 Mars Reconnaissance Orbiter took this image with its high-resolution camera, HIGHRISE. The heat shield had been dropped first, and landed furthest away from Curiosity. The backshell and parachute landed close together; apparently wind in the lower elevation was minimal. The descent stage had flown well away from the landing, to avoid disturbing Curiosity as it crashed (note the darkness around its impact area; residual fuel may have exploded). Source: NASA/JPL-Caltech/Un. of Arizona

The strategy for handling mass and momentum is the inverse between launching a rocket into space and recovering it from space to the surface of a planet or moon. To launch you have to give the payload the momentum and potential energy needed for its mission. To recover you have to dissipate the initial momentum and potential energy it brings from space. So we stage rockets to drop as much mass as we can on the way up and, similarly, drop as much mass as we can on the way down. The implications for a two-way mission are obvious. If you can use something on the way up that you used on the way down, that might help. (Image courtesy of NASA/JPL-Caltech.)

Results

Curiosity received some debris stirred up by the exhaust from the descent stage rockets, resulting in degradation of early pictures until the transparent lens cap was removed. Debris may have been responsible for the loss of a wind sensor. A future payload might be lightly shielded or longer cables used with the sky crane.

The time of landing was significantly longer than predicted after the descent stage started its final slow descent. The problem was in an overestimation (by 0.1%) of the value of gravity used in the descent stage computer and the pre-launch simulator. The value used was appropriate for the elevation of the landing site if had been on a flat plain, but the gravitational attraction of the masses beyond the crater's edge and Mount Sharp were not accounted for. As a result, the landing was softer than expected.

Although small, if the error had been in the other direction, the landing would have been harder and possibly overstressed Curiosity's suspension.

Curiosity was now a lander but not yet a rover. After doing some checks of the power, electrical, and thermal systems, initial camera images, before and after removing the transparent lens cap, were relayed through the Olympus orbiter, still in range. The next two days were used to replace the program used during transfer, entry, descent and landing with the surface operation program in each of the redundant A and B computers. Then Curiosity was a rover, ready for mobility checkout.

It was decided to make a strategic plan before doing a check out of the mobility modes. The first destination for investigation was to be an area called Yellowknife Bay, where there was a conjunction of three units of geology as selected from analysis of orbital photography and spectrometry. So the direction of motion was to the southeast, and different modes of mobility would be checked out and monitored, starting with short runs and gradually increasing them.

This was by far the most complex re-entry into a Solar System body from space that had been attempted. The reason for the combination of diverse technologies was that the atmosphere of Mars, which exists and must be dealt with but is not thick enough to be all that helpful. Each component of the MSL that participated in the entry, descent, and landing is scalable for the larger payloads that would be required for humans to explore Mars and return to Earth. We would, of course, require a scale-up of test facilities on Earth to support a manned mission to Mars in order to maintain or increase the reliability of this complex mode.

The success of this mode of entry, descent, and landing was a major contribution to Goal IV of the Mars Exploration Program—finding a way for humans to go to Mars and safely return to Earth.

Maps of the Landing Ellipse High resolution maps were prepared before the entry, descent, and landing to prepare for post-landing operations planning. Each map sheet, called a quadrant, covered a surface area of 1.2 km by 1.2 km. The 140 quadrants mapped all of the landing ellipse and additional context and traverse regions. They were prepared from orbital data, including high-resolution photography, elevation data, and geology. Further details on these maps are contained in an LPSC 2013 abstract by F. I. Calef III et al.

Each of these maps had a name, taken from a location on Earth. It turned out that the first planned goal after Bradbury Landing was located in the landing ellipse was an area whose quad had been named Yellowknife Bay. The strategic direction was to the West, to avoid the Bagnold dune field that would be a mobility problem on the way to Mount Sharp. However the diversion to Yellowknife Bay won out because a location that had been named Glenelg was an intersection of three distinct units of geology. The quad included an alluvial fan and probable ancient stream beds. As we will see in the next chapter, this decision delivered major science observations and set the pattern for the mission. The chosen path for Curiosity would be a compromise among scientific interest in a specific nearby location, safe mobility to reach it, and the strategic path to reach locations of Mount Sharp that had influenced the choice of Gale crater as the landing site.

Naming Another precedent set early in the operations mission relates to the problem of naming things. This included things like regions, mountains, valleys, stops of the rover, and even rocks of interest. Naming could have become a distracting problem for each day's planning meeting. The custom became to take names from a map of a location on Earth that had the same name as the quad being used for the location on Mars. This led to a ready list of choices for names. Of course, when Curiosity's path crossed a boundary between quadrants, the chosen names changed their style mysteriously.

Some of the names chosen had to do with similar geology, and some were confusing, mixing location names with geology concepts, and some were playful. But the point was, it was efficient, so why not? In any case, it got the job done without slowing down the work. And it was obviously fun.

4 Travel to a Dry Stream Bed and Yellow Knife Bay

The first days after landing were spent commissioning parts of Curiosity, but a number of pictures were taken and some science was done. This image from Bradbury Landing was taken on sol 2 from the left Navcam camera (see Fig. 4.1).

The rubble of pebbles and small sharp rocks in the foreground was revealed because the fine dust and sand was blown away by the exhaust from the rockets of the descent engines, leaving a darker surface. This rubble was a clue to two points that later became clearer: first, the pebbles and small rocks were more likely to have been carried here by flowing water than by winds and second, the sharp rocks, if obscured by dust and sand, could become a hazard to the rover

A few days later Curiosity turned her attention to the South and Mount Sharp. (see Fig. 4.2).

As Curiosity moved toward Yellowknife Bay in short increments, testing different modes of travel, further evidence of water transport was found—smooth pebbles too heavy to be moved by wind and water bound in minerals.

As the rover continued to check out different modes of mobility, an interesting area of rocks near a depression filled with sand appeared (see Fig. 4.3). This was taken as an opportunity to test the analysis instruments with the scoop instead of the more time-consuming drill and core sampling procedures.

After the scoop took the sample from the sandy surface, the "divot" (from the game of golf) shows the tendency to clump (see Fig. 4.4). The sand appears to be combined with fine silt that makes the bottom of the divot smooth, while the clump that has pulled away from the edge has a rough surface.

The CheMin X-ray diffraction instrument in the SAM sample analysis suite determined that the scoop sample had about a 1% content of water, combined in the crystalline grains of sand.

As well as taking a scoop sample of the sand patch at Rocknest, the Mastcam was used to do a similar analysis of a rock (see Fig. 4.5). Mastcam has three parts: a powerful laser to vaporize a small hole in the rock, a spectrograph to analyze the spectrum of the flash, and a micro-imager camera to show enlarged views of the holes in the rock.

Curiosity spent a month at Rocknest altogether, achieving many firsts (including many operational decisions to control all the parameters). She was so proud of how well her many instruments worked, as well as the planning and operational teams, that she took a selfie of herself (see Fig. 4.6).

The MAHLI, mounted on the turret at the end of the arm, was deliberately used to image as many visible surfaces of the rover as possible, so the resulting mosaic has some distortions and a cubist style. Selfies are particularly valuable to examine dust on surfaces, especially calibration targets for the cameras. Another important application is examination of the aluminum wheels for damage.

C. J. Byrne, *Travels with Curiosity*, https://doi.org/10.1007/978-3-030-53805-7_4

Fig. 4.1 An early image taken by Curiosity with her left Navcam on sol 2, the day after landing. The mountains are the northern edge of Gale crater. The shallow pits were left by the exhaust from the descent rocket engines. (Image courtesy of NASA/JPL-Caltech)

After the exploration of Rocknest, Curiosity continued the traverse toward Yellowknife Bay. From the arrival at Rocknest on sol 55 to the arrival at Yellowknife Bay on sol 130, Curiosity traveled a total of merely 200 m. This was a period of very intensive investigation of a very important area, the conjunction of three members of a formation that extends over a major part of the floor of Gale Crater. Figure 4.7 shows the relationship between this area and Bradbury Landing, and the inset shows the Yellowknife Bay in detail.

As the tour continues, special terms of geology and other sciences will appear and need explanation. Here we will talk about the terms formation, members of formations, and strata. Each of these is central to why this small area was so important to the selection of Gale Crater and the specific landing target for this mission.

Fig. 4.2 Early picture taken by Curiosity with her right Navcam on sol 12, showing Mount Sharp from Bradbury Landing. The dark band is the Bagnold Dunes that Curiosity must avoid to reach Mount Sharp. (Image courtesy of NASA/ JPL-Caltech)

Formation This very important term refers to the history of an extensive body of rock. As the name implies, the geologist wants to know how it was originally formed and modified. For example, the material may have come from a lava flow from a volcano or sediment from weathering of an earlier formation. It could have been transported by wind, water, or both. It may have been chemically modified by molecules of air in the atmosphere or by molecules dissolved in water. So a specific formation has a specific history. Two remote areas may have a specific history in terms of age or events and have the same formation name, sometimes derived from the location of the first formation.

Member A formation can have several members, each different from the others, as long as they were formed in the same way. Usually, the difference is expressed as an appearance, called a facies.

Fig. 4.3 Rocknest, in the Bradbury Group of geological formations. Curiosity took its first scoop sample in the sandy patch to the upper right and passed it on to SAM for analysis. The image has been white-balanced to show what the rocks and soils in it would look like if they were on Earth. (Image courtesy of NASA/JPL-Caltech)

Fig. 4.4 Curiosity's first scoop, at Rocknest. *Left*: Impression left in the sand after scoop has been lifted. *Right*: The sample in the scoop for sifting and transfer to SAM for analysis. These images are from the Mastcam instrument. (Image courtesy of NASA/JPL-Caltech/MSSS)

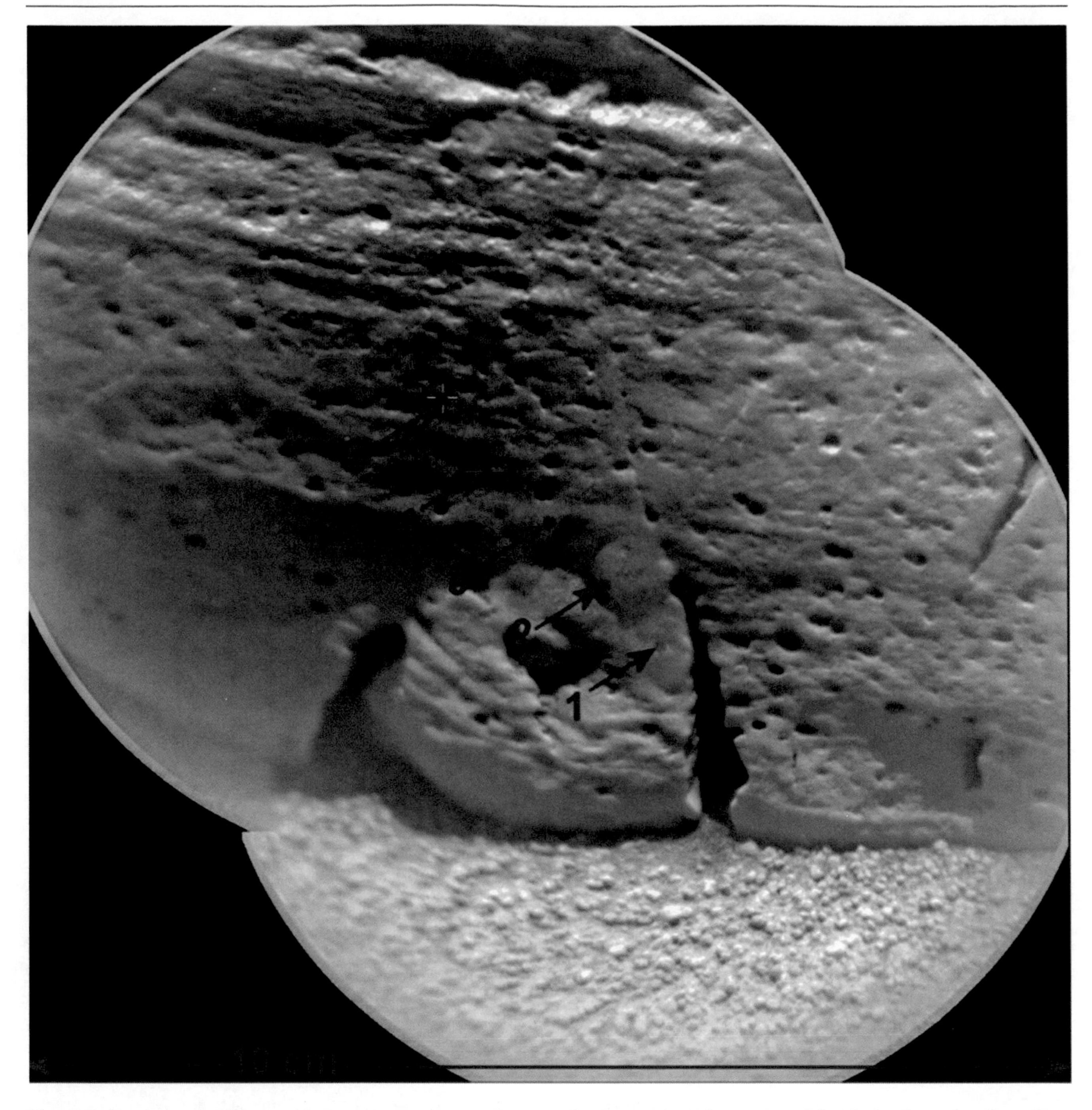

Fig. 4.5 Rocknest 3. This mosaic image superimposed on top shows a rock with a series of five holes from the ChemCam laser pulses. (Image courtesy of NASA/JPL-Caltech)

Fig. 4.6 Curiosity's first selfie, taken at the Rocknest site. This is a mosaic of images taken with the MAHLI camera, mounted on the arm. Of course, you cannot see the MAHLI or the end of the arm. There are four scoop divots in front of Curiosity. (Image courtesy of NASA/JPL-Caltech/MSSS)

Fig. 4.7 This is the path Curiosity took from Bradbury Landing to Yellowknife Bay by way of Rocknest, Point Lake, Glenelg, and Shaler. On this traverse various mobility modes were tested while moving toward Glenelg and Yellowknife Bay. At Rocknest, the first site exploration was performed, checking out planning and operations procedures as well as the science instruments and data communications. (Image courtesy of NASA/JPL-Caltech/MSSS)

Stratum (Plural: Strata) Formations usually occur in horizontal or near horizontal layers. For example, lava can flow down the slope of a volcano and pool on a plain at the bottom. Sediment can be transported to a lake by wind or water and settle on the bottom of the lake. Successive events can leave multiple layers of material, usually with the layer associated with a later event higher than a previous event. See Fig. 4.8 as an illustration of strata.

The view in Fig. 4.9 shows three different types of surface materials of different textures and brightness, one at upper left of the inset, one at the upper right, and one at the bottom. The different formations were observed from orbiting spacecraft. The ground view from Curiosity's camera is shown in Fig. 4.9. SPOILER ALERT: After reviewing all the evidence from the Yellowknife Bay area, these three converging bodies of rocks were deemed to be three members (Glenelg, Gillespie Lake, and Sheepbed) of the Yellowknife Bay formation.

Fig. 4.8 An illustration of a set of strata, exposed in a cliff. Each stratum of sediment was originally deposited by weathering of rocks of different chemical composition, resulting in the distinct color. Location: Argentina. The geologist, two meters from toes to fingers, is in the picture for scale. Source: Wikipedia web page labeled "strata"

Fig. 4.9 This view toward Point Lake shows the three types of intersecting material that were seen from orbit: Sheepbed, Gillespie Lake, and the dark material similar to that at Glenelg (X). At the top of the image is Gale Crater's rim. (Image courtesy of NASA/JPL-Caltech)

Fig. 4.10 The Shaler site contains thin rock slabs that appear to be of shale, a rock derived from silt and clay deposited in slow-moving water like lakes. Erosion, perhaps by wind or water, has undercut them by taking away lower layers of soft material. (Image courtesy of NASA/JPL-Caltech)

As Curiosity approached Point Lake, only about 20 m from Glenelg, the view from the ground was better than it would be at the specific point of Glenelg, so Curiosity observed the conjunction of the three members from the Point Lake location on sols 102 to 111.

After the rover left the Point Lake viewpoint, it was ready to tackle the Yellowknife Bay site, but first an interesting target showed up on sol 120. Shaler is an assortment of undercut slabs of rock exposures (see Fig. 4.10). As its name implies, it has the appearance of shale. An intensive study of these rocks was made with the ChemCam instrument when Curiosity returned to the site on sols 309 to 324. This type of rock is evidence of flowing water that deposited silt, eventually to become rock.

Once again, on to Yellowknife Bay! After the intense examination of the Rocknest area, the Point Lake viewpoint, and the encounter with Shaler, the next target was (finally) Yellowknife Bay. Low-angle high-resolution photographs from orbit suggested that a view from the eastern shore of the putative dry streambed might show clear strata. Indeed it did. See Fig. 4.11.

It is important to understand that all these names that were assigned in the course of Curiosity's mission were informal, subject to change as the ongoing flow of information changes the accepted paradigms. This applies to the categories of the named items, such as formation, member, and strata, as well as the names themselves.

Promoting Yellowknife Bay from an area to a formation as well as designating its members will have implications as the mission continues, at least in the floor of Gale crater.

Curiosity approached Yellowknife Bay from the east. It took very revealing pictures of the terrain across the dry streambed and drilled two core samples in the Gillespie Lake member, one at John Klein and one at Cumberland, a meter or two away. The Gillespie member may be an extension or the end of deposits from the alluvial fan of flow through the Peace Vallis, a valley in the rim of Gale crater.

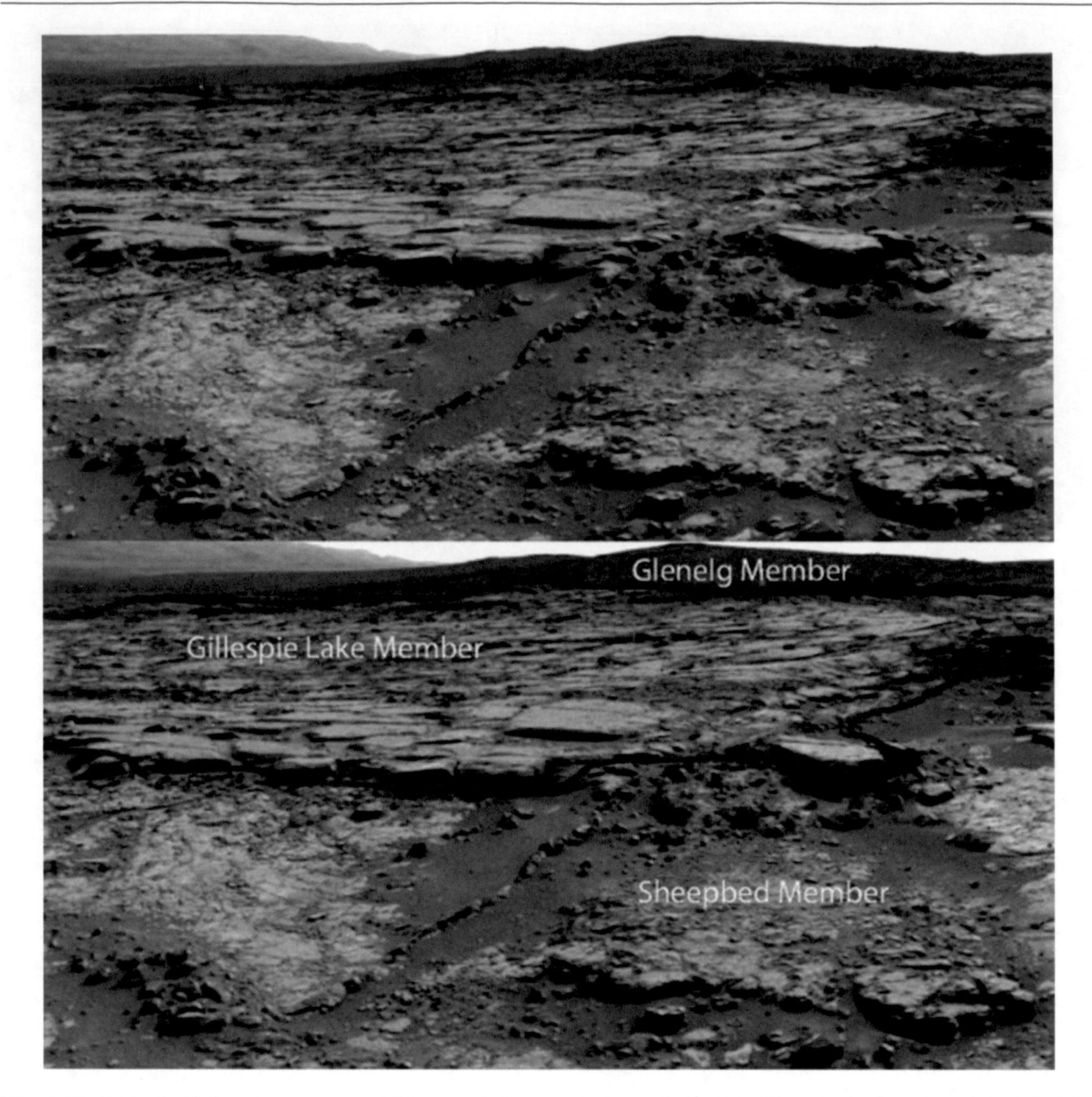

Fig. 4.11 Yellowknife Bay from the east. This remarkable mosaic from Curiosity's Mastcam images changed minds. The name "Glenelg" had been assigned to a conjunction of three possible separate formations. Now, after intensive investigation on site (in situ, if you prefer), the name is assigned to the Glenelg Member of the Yellowknife Bay Formation. (Image courtesy of NASA/JPL-Caltech/MSSS)

Fig. 4.12 This is the base rock of the streambed at location John Klein. A, B, and C are enlargements of rocks that have been carried along the ancient stream and indicate its depth and flow rate. (Image courtesy of NASA/JPL-Caltech)

Fig. 4.13 The tailings are from rhe top of the John Klein drill hole and the sample (see Fig. 4.14) has been lifted out. (Image courtesy of NASA/JPL-Caltech)

Fig. 4.14 The sample from the John Klein drill hole has been lifted to the tray by the scoop, for seiving and transport to the SAM suite for analysis. For this picture, MAHLI was brought close to the sample in the tray and the focus set accordingly. This made the background surface out of focus. (Image courtesy of NASA/JPL-Caltech)

As the climate dried, the last probable flow would have been in the Sheepbed member, although it is possible that unit was much older than the last flow of water, and wind or water flow erosion has exposed it.

The John Klein and Cumberland samples from drilled holes were taken from adjacent exposed rocks in the probable stream of flowing water (see Fig. 4.12). The resulting SAM analysis confirmed that the rock was formed by sediment in water, and provided evidence of the nature of that water Fig. 4.13.

After examining Cumberland with the contact instruments and drilling and transferring the core sample to the SAM instrument suite, Curiosity started her retreat from the Yellowknife Bay area. As she left, she revisited earlier sites to take further data suggested by the analysis of earlier observation. she started the retreat on sol 295, traveling back toward Point Lake to examine a puzzling outcrop there. After revisiting Slater, Curiosity began the long travel along the Bagnold Dunes to reach a narrow crossing point for the ascent of Mount Sharp.

The diversion from Mount Sharp to Yellowknife Bay resulted in meeting the mission goals in a relatively short time because of the prudent decision to avoid an early attempt to cross the Bagnold Dunes. Indeed, the broader goals of NASA's Mars Exploration Program Goals received major contributions, like the following:

Results

Goal I: Mars Habitability

The rocks on the floor of Gale Crater in the Yellowknife Bay area were formed in the presence of water, especially the mudrock portion. And mudstone means there was a lake.

The instruments in the SAM suite of instruments analyzed the samples of the Sheepbed unit of the Yellowknife Bay formation and found that it was principally mudstone. The percent of mudstone could be detected in other rocks as well by the APSX instrument, which examines crystaline minerals by X-ray diffraction. The percentage of the examined rock that is amorphous (non-crystaline) is found to be mudstone. The mudstone component in other units is similar in element percentages, as found by the Mastcam instrument, that measures the elements by the light from its laser flash.

Mudstone is the geological name of a class of a mineral similar to shale and siltstone, which differ in the ratio of components and the grain size of the amorphose component. If the dominant component in a rock is amorphous, that is mudstone, then the rock is called mudstorne. The rocks in the nearby outcrop Shaler have the character of shale, easily split into sheets because of their clay content as well as mudstone. The outcrop is called Shaler after a small town in Pennsylvania.

Mudstone is formed by sediment from a body of slow-flowing water like a lake, which carries fine grains of minerals that settle to the bottom and are lithified (made into rock) by pressure of higher layers and by dissolved chemicals that form a cement to lock the grains together. So the geologists concluded that the floor of Gale Crater was once the bottom of a large lake long ago.

Flowing Water Visual observation from the many cameras showed clear evidence of streams with at least a meter or two of water flowing rapidly enough to move rocks of at least tens of centimeters. So eventually the lake, in the process of drying up, must have been reduced to flowing streams in the Gillespie Lake member. Other evidence observed by Curisosity's instruments, including the SAM suite, establish that the flowing water had a neutral Ph and low salinity.

Elements and Energy for Life SAM, CheMin, ChemCam, and AXSM together established that the key elements needed for life as we know it in bacteria have been found in the Yellowknife formation. These are carbon, hydrogen, oxygen, sulfur, nitrogen, and potassium.

CheMin has found that Iron and Sulfur are found in minerals with difference valences (redox states). This implies that some of the minerals were formed in different environments. This means that there is available chemical energy for life below the surface, protected from the solar wind, if it has evolved with an appropriate metabolism.

In summary, Curiosity's exploration of the Yellowknife formation had resulted in enough findings by sol 323, having driven 848 m from Bradbury Landing, to meet its goal of having determined whether the Mars envioroment could once have supported life. The answer is yes. Life has not been detected on the surface or in the samples, a few cm in depth, but the search will continue in the ensuing mission.

How long could Mars have supported life? The ongoing mission to Mount Sharp may contribute to answering that question. The surfaces on Mount Sharp are laminated, implying a long period of formation. As we climb higher we may be able to establish whether those layers were formed in life-friendly environments. Assuming that life started spontaneously on Earth, it may have had as much as 700 million years to have occurred by chance. Of course, it may have started earlier.

Absolute ages may await samples returned to Earth. Proving that life never evolved on Mars, like any negative proposition, would be hard. It would require deep drilling, sample return to Earth, and exploration of many other areas of Mars, even the polar areas.

As planned when Curiosity was still at Bradbury Landing, Curiosity would proceed from the exploration of the Yellowknife area westward to lower Mount Sharp. It would avoid crossing the Bagnold Dunes until a safe place could be found for mobility. The planned route can be seen in Fig. 5.1.

Methane On June 15, 2013 (sol 305), Curiosity ran an environmental test while it was between revisits to Point Lake and Shaler and as it was preparing to leave the Yellowknife Bay area for the traverse southwest along the Bagnold dune field to Mount Sharp. A sample of the Mars atamosphere was brought into the SAM instrument suite for analysis and found methane Methane is a very simple carbohydrate, one atom of carbon bound to four of hydrogen, detectable by SAM at very low levels. On Earth, methane (the major component of natural gas and a common waste product of metabolism) is a very significant component of Earth's atmosphere.

Although such environmental samples had been taken and analyzed several times previously in the 305 sols since Curiosity landed, this was the first time a significant trace of methane was found: six parts in a billion by volume (the background measurements were 0.3 parts per billion). Previous measurements from Earth and from Mars satellites had detected ocasional sporadic measurements of methane, but none were confirmed. After nearly a year of no significant methane data from Curiosity's SAM, the responsible scientists for this controversial measurement published a paper reporting their negative result. The technical news media announced "Mars has no methane," implying "Mars has no life."

The new discovery on sol 305 of Curiosity's mission was thus major news, but of course subject to question. Two sattelites in Mars orbits with the capability for trace gas detection did not confirm the Curiosity event. Was it just another suspect datum point? Terrestrial contamination? A few random molecules inorganically generated? So, why not wait a while and see if it happens again.

It turns out that it did. But that is a story for later chapters, starting with the next.

Note A new method of analysis of archival Mars Express data on June 16, 2013 (yes, the next day after the Curiosity methane event) detected methane in the Mars atmosphere. This was published in June of 2019. More news in a later chapter.

Goal III: Geology

By exploring the common area of three different regions, as seen from orbit, it has been possible to characterize the Yellowknife Bay formation, consisting of three members, the Glenelg member, the Point Gillespie member, and the Sheepbed member. Although they are informal names, they will be used as such until approved or replaced by standard terms. The Yellowknife Bay formation might be used to characterize complex deposits near the end of alluvial fans such as were formed by flow over the northern rim of Gale crater through Peace Valley. The Glenelg member appears to be very large, extending far to the west, and may represent a layer that once covered the Lake Gillepsie and Sheepbed members and then eroded away. The bottom (oldest) Sheepbed member may represent an early slow flow through a break in the rim of Gale crater, perhaps being filtered through the material of the crater wall before Peace Vallis was enlarged. The Sheepbed formation might be applied to any mudstone deposit of similar composition that might underlie the other members of Yellowknife Bay over part of their area. Bradbury Group is being used to encompose multiple formations at the floor of Gale crater, in parts of the region formally called Aeolis Palus.

5 Travel Along the Bagnold Dunes to Mount Sharp

The science team agreed with the engineering team to plan for a least-time route to a point where it was safe to cross the Bagnold Dunes (see Figs. 5.1 and 5.2). The target for the traverse was the Pahrump Hills, considered to be the foothills of Mount Sharp and not part of Aeolis Palus, the floor of Gale crater.

Way stops were planned for remote science at Darwin and Cooperstown. One stop for contact science was planned for the Kimberly area because of interest in the geology there.

Soft, deep sand is hazardous for rovers. Spirit's mobility was ended when it entered such an area. Opportunity was held up for a month. Even Curiosity encountered a problem further along as it crossed the Bagnold Dunes at their narrowest point. It is easy to see why by looking at Fig. 5.2.

Darwin was the first way stop on the traverse to the entry to Mount Sharp on sol 392, after traveling 2,852 m. Curiosity stopped about a 100 m short of Darwin to take the remarkable panorama of the foothills of Mount Sharp, seen in Fig. 5.3. The scene extends from the mounds and valleys of the foothills near Darwin to the higher, rugged hill to the west. The mounds show clear laminations, a record of separate formation or erosion events that offer clues to when they formed. The sandy dunes in the lower part of the image look very like a river. Once, perhaps as a lake dried up, there may have been a river or pond there, in a depression. Now, the sand settles there as it does in all low areas, deposited by the wind. The surface of the loose sand, blown by the wind, can appear like waves in water.

Between Darwin and Cooperstown, at sol 426, the surface, evaluated from orbital images, changed to rugged terrain, a marked contrast to the hummocky plains of Bradbury Landing, the Glenelg member of the Yellowknife Bay formation, and Darwin.

Figure 5.4 shows the approach to Cooperstown, with rocks scattered on the path to an outcrop of rocks along a ridge. After examining these rocks on sol 442 with a MAHLI panorama, Curiosity proceeded toward Kimberly.

The Kimberly area was chosen for contact science examination because of an apparent history of water flow (see Fig. 5.5) and for three interesting mounds in the area. One of these mounds, called Mount Remarkable, can be seen in Fig. 5.6. The Kimberly on Mars was named after "the Kimberly" in Australia, a district (not a town) of the state of Western Australia. A local custom in Australia is to refer to "the Kimberly" district and not to use "Kimberly" alone. The custom has been to transferred to the Kimberly, Mars.

Mount Remarkable is about 5 m high. By examining a layer of rock exposed at its base, Curiosity gained information about similar rock 5 m below its peak without drilling. Similar methods are used by geologists on Earth to examine road cuts to infer properties of deep rock layers nearby. A selfie of Curiosity examining a rock called Stephan at the base of Mount Remarkable with its Mastcam is shown in Fig. 5.7.

C. J. Byrne, *Travels with Curiosity*, https://doi.org/10.1007/978-3-030-53805-7_5

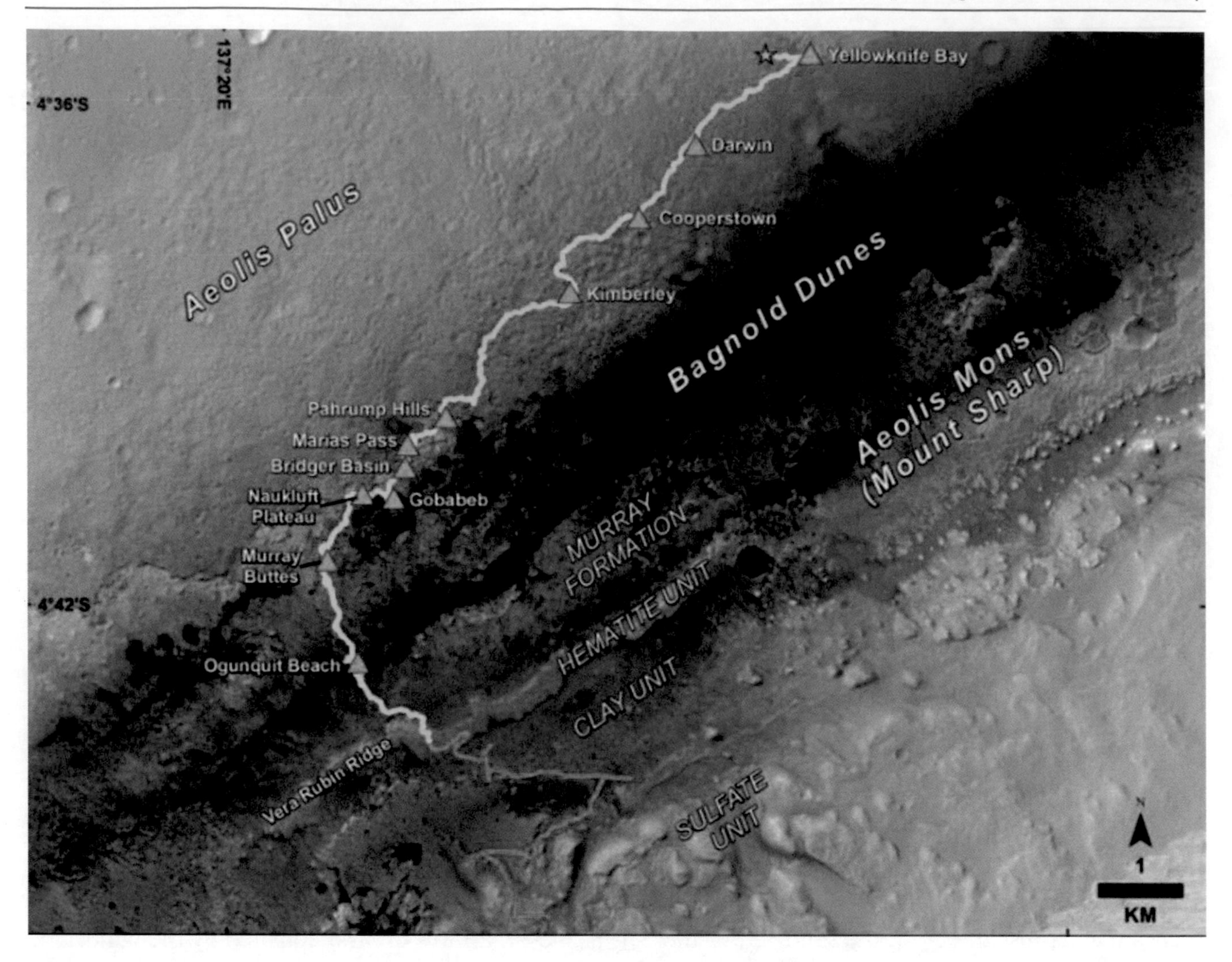

Fig. 5.1 This shows the route of the traverse from the Yellowknife Bay area to the Pahrump Hills, the foothills of Mount Sharp. In yellow, the route looks irregular because it was planned to be as level as possible and relatively free of sharp rocks that could be a hazard to the wheels. The strategic plan was modified by a tactical plan as needed. (Image courtesy of NASA/JPL-Caltech)

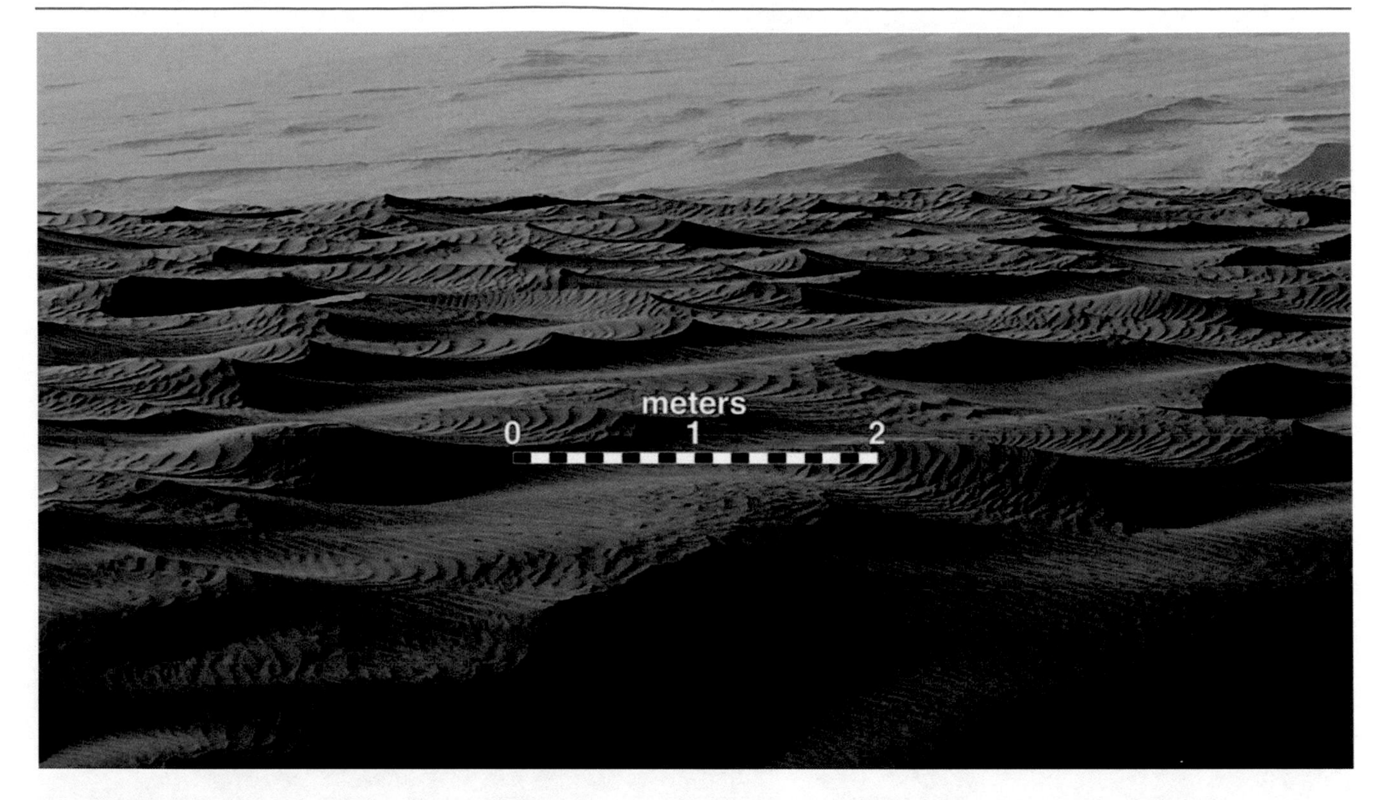

Fig. 5.2 Part of the long field of Bagnold Dunes. It is easy to see that the wheels could lose traction if the base or suspension of Curiosity got hung up on a ridge. Crossing a field of such dunes would at best be a slow procedure. (Image courtesy of NASA/JPL-Caltech)

Fig. 5.3 This panorama was constructed from images taken by the Mastcam about100 m east of way stop Darwin on sol 387. The image has been color-balanced to show the rock tones as they would appear on Earth. That process has also turned the Mars sky (upper right corner) from red to blue. (Image courtesy of NASA/JPL-Caltech/MSSS)

Fig. 5.4 Cooperstown. As Curiosity continued along the Bagnold Dunes toward an entrance to Mount Sharp, the terrain became much rougher. This mosaic shows rocks scattered along a ridge. (Image courtesy of NASA/JPL-Caltech)

Fig. 5.5 This scene at Kimberly justified its selection for extensive study with the contact instruments. The long slope to a large depression suggested water flow. The DAN instrument detected a high level of hydrogen, most likely locked into the crystalline mineral structure of the rock. The rover's elevation here is about 20 m higher than that of Bradbury Landing. (Image courtesy of NASA/JPL-Caltech)

Fig. 5.6 Mount Remarkable in the Kimberly area. The base of the left slope of Mount Remarkable can be seen at the right edge of the close-up view in Fig. 5.5. (Image courtesy of NASA/JPL-Caltech/Ken Kremer/Marco Di Lorenzo, from apod.nasa.gov)

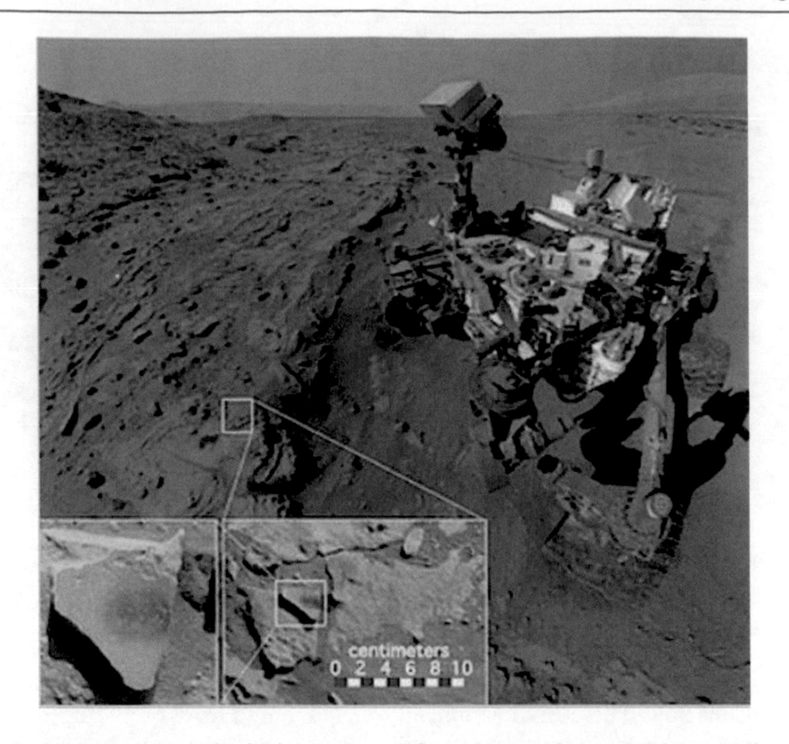

Fig. 5.7 A bit of virtuoso imagery on Mars. Curiosity took a selfie as it was about to zap a small rock called Stephen with its ChemCam's laser. The middle inset shows the magnified rock with an overlaid scale bar. The left inset shows the target rock magnified again. The rock had been dusted with the brush on the turret, showing the rock's color without the dust layer. The turret then brought the MAHLI in position to make the selfie mosaic. The arm did not bring MAHLI to Curiosity's right side to avoid disturbing the target rock. Then ChemCam shot the laser and created a digital spectrogram of the flash. The background may have been photographed before Curiosity came around to take its position. (Image courtesy of NASA/JPL-Caltech/MSSS)

A core sample was taken at Windjana, a larger bedrock area next to Stephen, in the Kimberly area on sol 621 (Fig. 5.8). Like Stephen, the rock was found to contain a high percentage of manganese oxide, with implications about the environment of its formation, described in the "Results" section of this chapter. The sample was transferred to SAM for analysis.

Curiosity left the Kimberly area on sol 630, traveling toward the entrance to Mount Sharp. Although alternate plans were reviewed, the next area for contact science turned out to be the Pahrump Hills at the base of Mount Sharp, beyond Aeolis Palus (the floor of Gale crater). The Pahrump Hills are short of the Bagnold Dune field, but there are opportunities for passing the dunes beyond the Pahrump Hills.

Fig. 5.8 Curiosity is parked by a sandstone rock called Windjana that she is about to drill for a core sample. In Western Australia, there is a national park in the Kimberly district. The arm is stowed. (Image courtesy of NASA/JPL-Caltech)

Results

Here are some results since leaving the Yellowknife Bay area:

Methane Measurements A second set of significantly high atmospheric methane levels was reported starting at sol 466 and ending at sol 525 (see Fig. 5.9). In this interval, Curiosity was traveling between Cooperstown and Kimberly. The earlier high level at sol 305 was taken when Curiosity was finishing its exploration of Yellowknife Bay, between Point Lake and Shaler.

Apparently, methane is released into the atmosphere sporadically, in "burps," as it is said. It does not accumulate in the atmosphere of Mars. Mars lacks an ozone layer in its atmosphere. There is a high level of ultraviolet radiation from the Sun, stimulating methane to react with oxygen, producing carbon dioxide and water but the rate of decay is unexpectedly high.

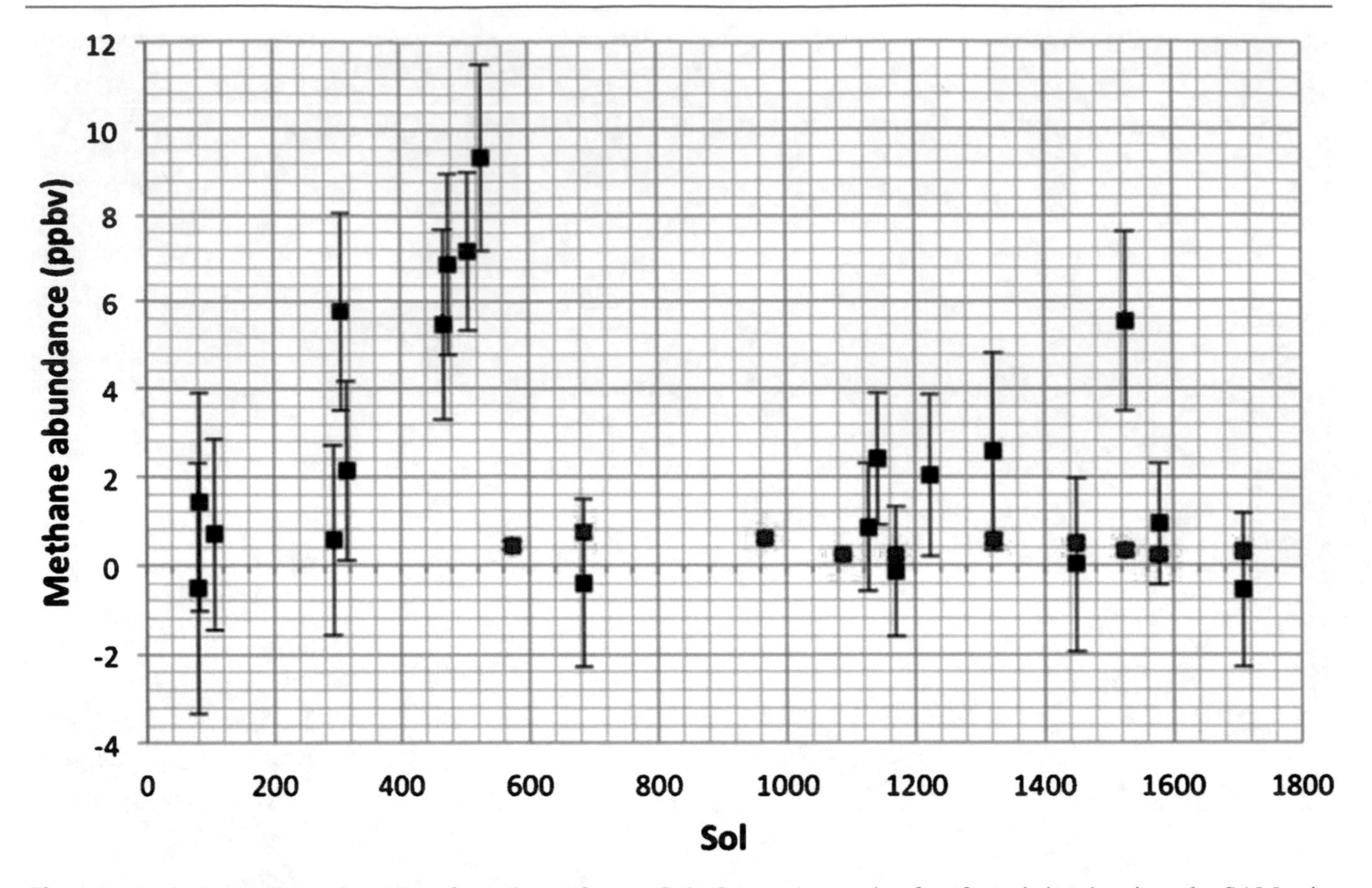

Fig. 5.9 Curiosity methane data taken from the surface at Gale Crater. A sample of surface air is taken into the SAM suite of instruments and measured. The data points with error bars are the direct measurement. The red points relate to an optional process to remove carbon dioxide and water vapor so that methane and other components are enhanced for more accurate measurement. (Figure courtesy of Jorge Pla-Garcia et al., JGR: Planets, Volume 124, Issue 8, April 24, 2019 (open access))

Although Mars Express (ESA) and MAVEN (NASA) have been in orbit, and can detect methane, they did not detect a sporadic bump in methane at the time of the two events shown in Fig. 5.9. So Curiosity's detection, like those from orbit and telescopes from Earth, joined in contributing unconfirmed sporadic events. Usually, we have been following the path of Curiosity describing events as they happened along the way. In the case of methane the story is still evolving.

Note that there are data points in the graph of Fig. 5.9 that are labeled red for "Enrichment." This is a mode of the SAM instrument suite that passes an atmospheric sample through filters to remove carbon dioxide and water vapor, the largest components of the atmosphere of Mars. The remaining components are accumulated to fill the volume of the sample container. This increases the sensitivity of the analysis for all of the remaining gas by a factor of 30. Enrichment samples take much more time and power, but the decision was made to take them to see if there were more events to examine. It turns out that there were (see Fig. 5.10). Recently, in June, 2019, SAM reported a record methane spike of 21 parts per billion on June 22, 2019 (sol 2442) (CNET, June 24, 2019) https://www.cnet.com/news/nasa-confirms-curiosity-rovers-surprising-mars-methane-discovery/. A few days later, the methane had returned to background level. However, in this case, the spike was confirmed from orbit, centered near Gale crater in an area of linear faults in the Aeolis region.

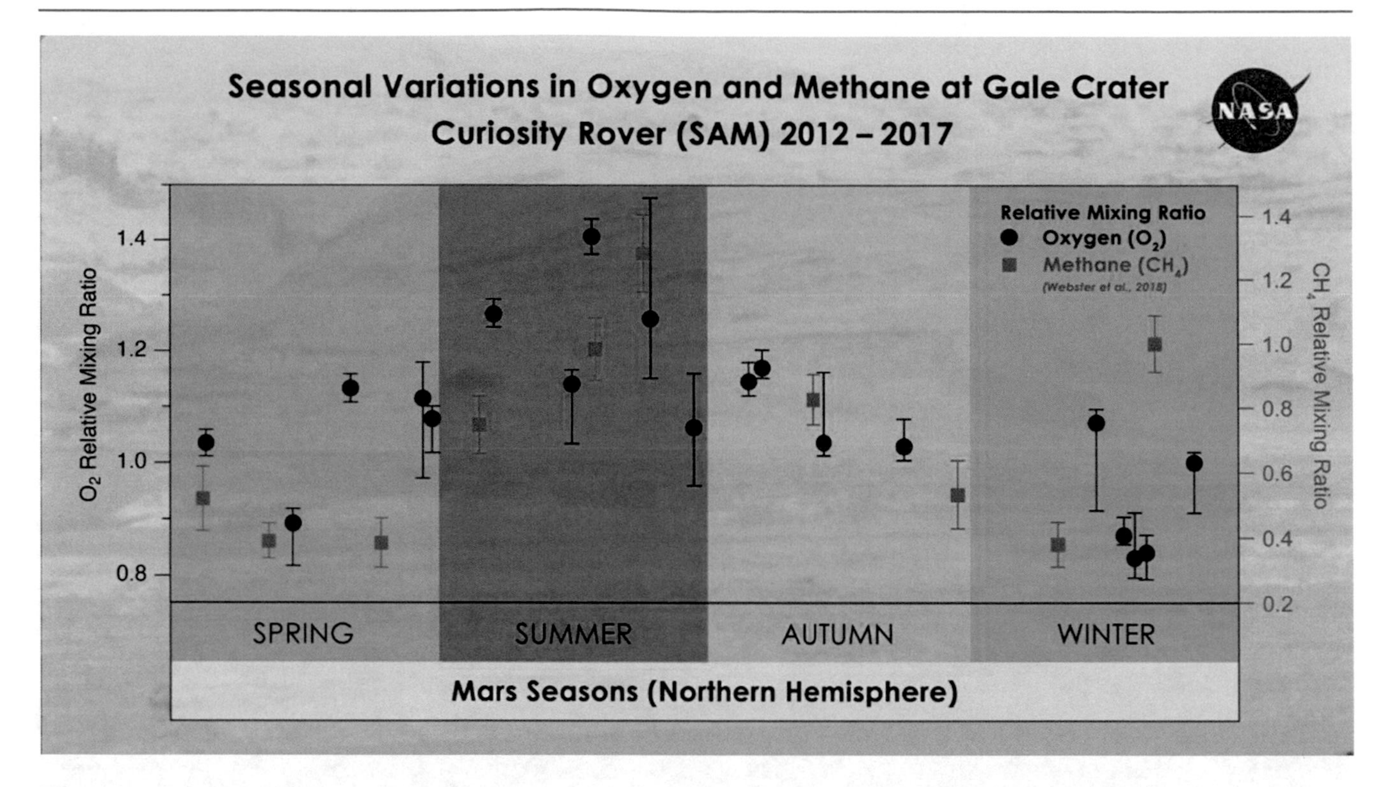

Fig. 5.10 Mars methane and oxygen by season. Curiosity sampled the local Mars atmosphere over three Mars years, using the enrichment mode (see text). Plotted by the local season, (solar longitude). the background level of methane rises at the surface of Gale Crater (near the equator), heats and drops as the surface cools. The maximum surface temper is in spring because Mars is closer to the sun in late spring in this time frame. There is a time lag between surface temperature and relative abundance, suggesting that heat must take time to travel from the surface down to the source of the plumes of methane. See the results section of the chapter on the Clay-Bearing unit, where Curiosity was when she reported the record spike in methane concentration. (Courtesy NASA/Goddard/Dan Gallagher/Melissa Trainer.) downloaded from https://mars.nasa.gov/news/8548/

Seasonal Variation In the enrichment mode, several measurements per year were made, with significant results. The background detection of methane was not instrumental error but showed a seasonal variation (see Fig. 5.10). Mars has a polar axis tilt of about the same angle as Earth relative to the plane of the Solar System, so even though Gale crater is near the equator of Mars it has seasons. Because Mars has an elliptic orbit around the Sun, there is an additional thermal pattern added to the one caused by the axial tilt. Each of the two seasonal cycles has a period of about two Earth years.

The seasonal variation suggested an insight into why methane was so elusive. Perhaps when the surface is heated, a burst of methane is released from the surface rocks or from some depth below the surface. Such thoughts prompted a study of how methane could be generated, stored, released and removed (see Fig. 5.11). On Earth, methane primarily comes from the metabolism process or the decay of life. On both Earth and Mars, it can also come from the interaction of some rocks such as olivine and water. However generated, it tends to rise to the surface but can be blocked by being frozen, mixed with water. Such a state is known as a clathrate. If the temperature rises, it can be released. This process could explain the randomness of when and where it is released. After release, methane can be removed from the Mars atmosphere by ultraviolet rays from the Sun, leaving carbon dioxide and water vapor in its place.

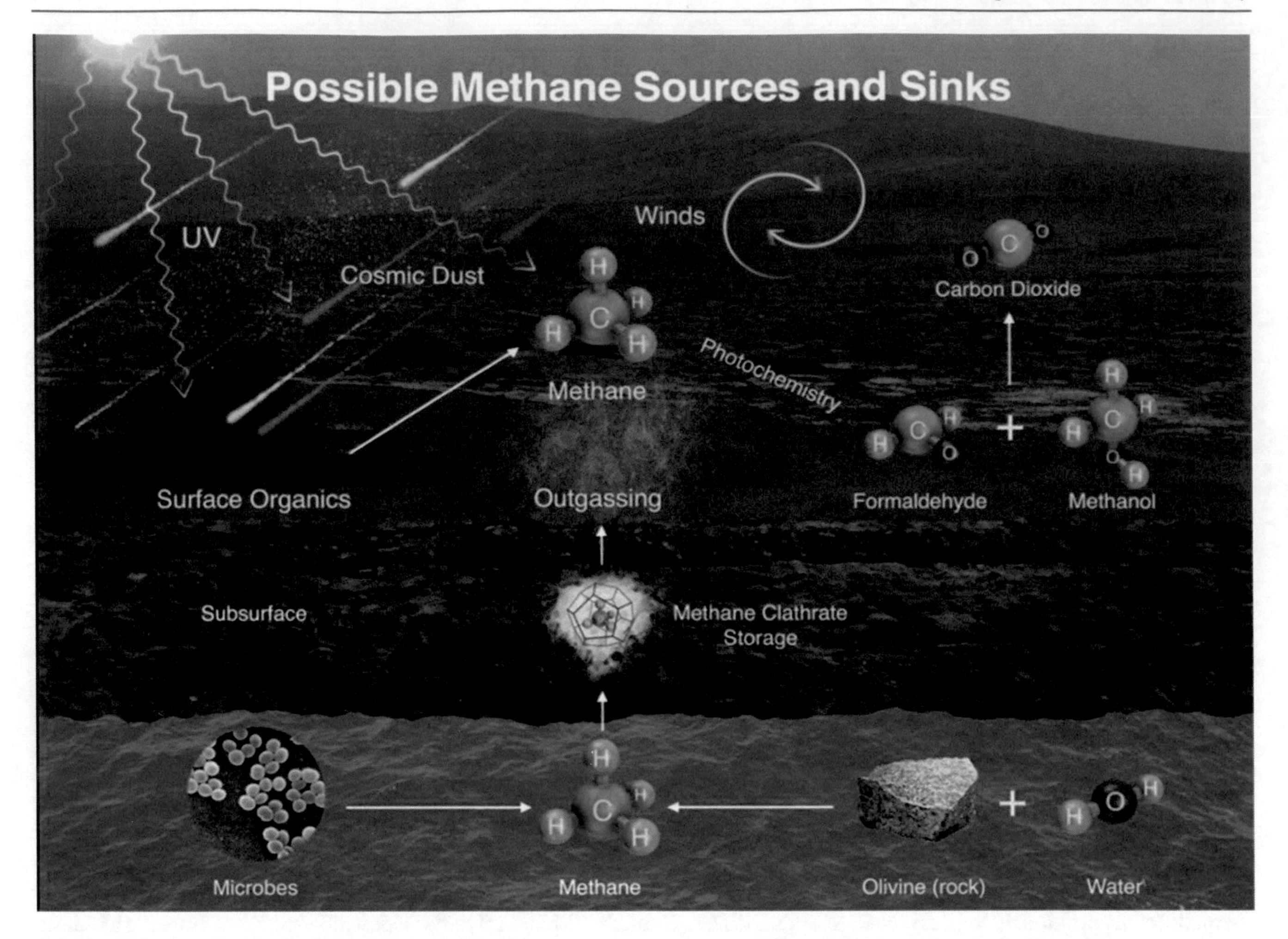

Fig. 5.11 Flow of methane in the Mars atmosphere. (Image courtesy of NASA/JPL-Caltech/msl/pia19088)

That was the thinking at the end of 2018. The erratic measurements of methane (or not) could be caused by local heating of the mix of methane and water (clathrate) frozen below the surface. In January of 2019, a new finding was announced by ESA's Mars Express orbiter team. Analyzing archival data of Mars Express, the team was able to detect methane in the atmosphere near Gale crater a day after Curiosity detected it (June 13, 2013) at the surface. Mars Express detected the methane in an area just to the southeast of Gale crater (see Fig. 5.12). The area has a number of linear features that could be faults, which have the potential of releasing methane from below the surface.

This remarkable verification of simultaneous observation from the surface and orbit confirms that the release of methane at Mars takes place at random times and unexpected localities but is real. It is just hard to find because you have to look in the right place at the right time.

Another discovery was made in 2019 after the enrichment-mode seasonal data for oxygen was graphed, including the data taken during the traverse along the Bagnold Dunes (see Fig. 5.10). The oxygen had the same seasonal behavior as the methane! This led to a new hypothesis that the joint cause was that at least partially due to seasonal freezing of carbon dioxide during the South Pole summer, reduced its amount in the atmosphere. Thereby the percentage of oxygen and methane in the atmosphere would be increased.

Manganese Oxide The SAM analysis of the sample at the base of Mount Remarkable showed a high level of manganese oxide, which is quite rare. Its presence indicates it was formed in an environment where oxygen was present in the atmosphere at some time in the past.

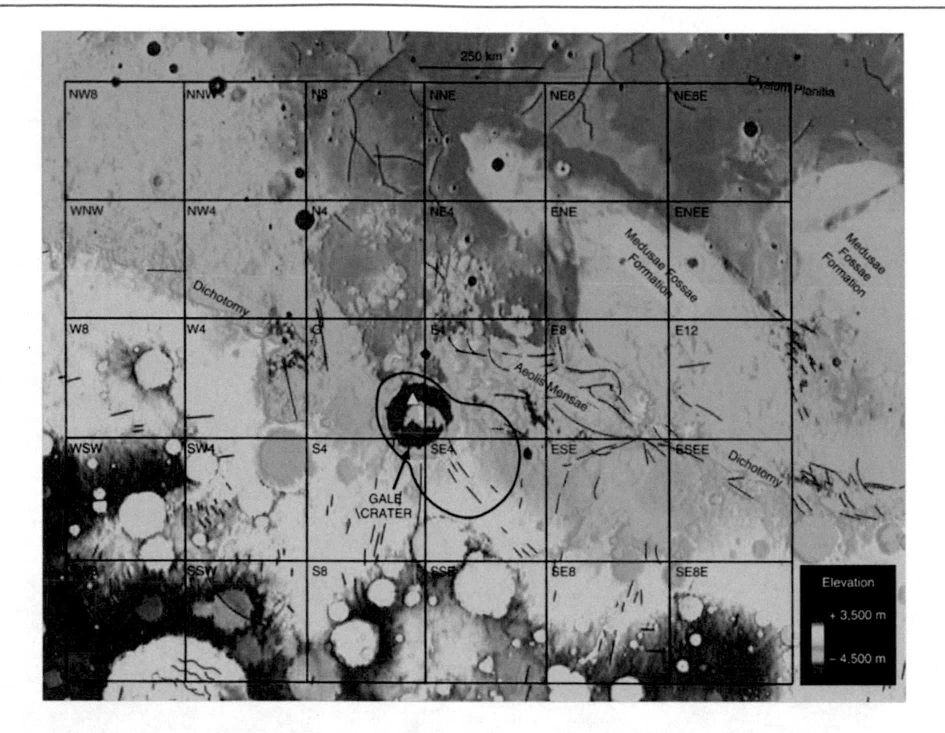

Fig. 5.12 Methane confirmation, orbit to surface. Mars Express archival data confirms a methane spike near Gale Crater. The large black loop is a probable area of a methane plume. Red lines in the topographic base map of the area were identified as extensional faults. (Image courtesy of ESA/Giuranna et al, Independent confirmation of a methane spike on Mars and a source region east of Gale crater, Nature Geoscience 12, 2019. Base topographic map, NASA/MOLA/Goddard/USGS)

DAN Results As Curiosity roved the Kimberly area, the DAN instrument recorded that its neutron beam return indicated a relatively high degree of hydrogen atoms within a meter of the surface, probably water bound in clay or other minerals.

Mobility There is good news and bad news and more bad news. In response to everyone's wish to get on to the slopes of Mount Sharp, and with good experience in mobility at Yellowknife Bay, the navigation planners made excellent progress with long daily drives, fast autonomous drives, and two-day drives, where Curiosity stops for a night and then continues the next day from the same point.

The first bad news occurred when it passed a boundary between the hummocky plains it traveled since leaving Slater and the rugged terrain it was in after sol 426. When a survey of the wheels was done on sol 463 with the MAHLI hand lens, the wheels showed quite a bit of damage, possibly in that 37 days of driving in the rugged terrain.

The rugged terrain had an unusual surface. There were lots of scattered rocks, left over from the removal of layers of rock by wind and sand erosion. They were partly buried in sand, leaving edges sticking out of the sand to be polished into points by windy sandblasting. The weight of the rover drove the points into the aluminum wheels, punching holes and causing cracks. As a result, shorter drives were planned, and the worst rocky areas were avoided if possible, lengthening the path. Naturally, progress was slowed as it traveled to the interesting formations of lower Mount Sharp. Fortunately, the new strategy for limiting damage to the wheels was found to be successful. Fig. 5.13 shows moderate damage to the wheel surface, about thirty-thousandths of an inch thick) when Curiosity was at Kimberly.

Later, in 2019, by the time the rover was at the Vera Rubin Ridge, two grousers (raised quarter-inch treads) had broken.

Fig. 5.13 Wheel damage as of April 18, 2016, sol 1315, was moderate in this MAHLI image of the left middle and rear wheels. (Image courtesy of NASA/JPL-Caltech/MSSS)

6 The Pahrump Hills, Lake Sediment, and the Murray Formation

When do you get to the bottom of a mountain? The top of a mountain is clearly marked—the peak. One answer is you get to the terrain that is similar to that on the mountain but distinctly different from the surrounding plain. In the case of Mount Sharp (Mons Aeolis) the bottom is the intersection of the crater floor (Aeolis Palus) with the upward slope and foothills of Mount Sharp. Curiosity reached Mount Sharp on September 14, 2014, on sol 751 (751 Martian days after touchdown at Bradbury Landing). The way stop is named Piute, in the Amargosa Valley.

After Piute, a short drive brought Curiosity to the beginning of the Pahrump Hills campaign at the Confidence Hills way stop. Although the Bagnold dune field is still between the Pahrump Hills and the peak of Mount Sharp, there is a path with reasonable mobility that leads around two large dunes to the Murray Buttes and beyond. The Murray Buttes were identified from a review of orbital photography as a major target for geology and also promises spectacular images.

Curiosity reached the Pahrump Hills on September 18, 2014 (see Figs. 6.1 and 6.2) on sol 753, after traveling 6.9 km from the Shaler outcrop in the Yellowknife Bay area. After careful exploration, sampling and analysis, the Murray formation was defined and Pahrump Hills became the defining site for its first geologic member.

Murray formation rocks were found to be deposited by liquid water. They consist of fine- and coarse-grained diverse components dominated by mudstone. The facies (face surface) of the bedrock of the Pahrump Hills member of the Murray formation is flat, light-colored (after dust is brushed away) and fractured.

The Murray formation is the lower strata on the north side of Mount Sharp. The upper limit remains uncertain as Curiosity continues to climb, but is at least 300 m above the base. As the exploration of Mount Sharp has continued, the definition of the Murray formation and its members has been both broadened and become more limited as new observations have been added.

Curiosity entered the Pahrump Hills area through the Amargosa Valley, reaching the Pahrump Hills at the Confidence Hills site (see Figs. 6.1 and 6.2). There, it took the first core sample since Kimberly and passed it to SAM. It then did a walk-about of the Pahrump Hills area (see Fig. 6.3). A walk-about is what geologists do when they enter a new area. They walk around the area to get the general context of the region, identifying points of interest for further observation and planning a sequence of detailed remote and contact examination and sampling.

Curiosity, for its walk-about, followed a meandering path on many drives, being directed to approach diverse outcrops to get a closer view of the rocks with the remote instruments (see Fig. 6.3).

At the end of the remote instrument walk-about they took a very good look at Whale Rock (Figs. 6.4 and 6.5), which displays many fine striations, indicating separate flooding events that deposited sediments. These striations occur throughout lower Mount Sharp and are an indication of the passage of time.

C. J. Byrne, *Travels with Curiosity*, https://doi.org/10.1007/978-3-030-53805-7_6

Fig. 6.1 Curiosity took this image of the Pahrump Hills area with a front Hazcam on sol 752 as it was about to leave the Amargosa Valley and approach the Confidence Hills drill site. Mount Sharp is at the horizon. (Image courtesy of NASA/JPL-Caltech)

After reviewing the data from the walk-about, the science team selected targets for contact instruments, at locations shown at the red dots, to examine the grain size distribution and element abundance of the rocks. Then, reviewing that data, they selected locations for drilling and collecting the three core samples at Confidence Hills, Mojave 2 and Telegraph Peak. From SAM data, the mineral structure of the rocks was revealed, complementing the chemical data from the contact instruments. So in covering this area, the cautious strategy implied multiple traverses to gain the most knowledge in a responsible way. This strategy was followed repeatedly at selected science-intensive areas throughout the mission—walk through with remote instruments, contact science, then samples.

After the walk-about, Curiosity revisited some other outcrops of rock and returned to the Confidence Hills way stop for further examination and also took its first drill sample of three in the Pahrump Hills (Fig. 6.5).

Fig. 6.2 Curiosity at Pahrump Hills, at the base of Mount Sharp. Curiosity is inside the blue square. (Image courtesy of NASA/JPL-Caltech/MRO/MOLA)

Fig. 6.3 Curiosity entered the Pahrump Hills at the base of Mount Sharp from the north at Confidence Hills. This view is toward the southwest. The Bagnold Dune field is the dark area at the upper left of the image. Beyond the dunes are higher foothills of Mount Sharp. The yellow path marks the first walk-about. Blue dots mark drill sample sites. (Image courtesy of NASA/JPL-Caltech)

The next contact site for Pahrump Hills after Confidence Hills was at the Mojave way stop. (See Fig. 6.6 for a selfie of Curiosity and a view of the site.) Views of Whale rock are at Fig. 6.4 and Fig. 6.5. After Curiosity left Wale Rock, it was decided to drill for a sample at Mojave.

Fig. 6.4 Whale Rock is a large outcrop in the Pahrump Hills area. It is at a relatively high altitude in the area, near the highest point visited by Curiosity, Telegraph Peak. Its striations are an indication of the ages and are a forerunner of many such exposures throughout Curiosity's path on lower Mount Sharp. (Image courtesy of NASA/JPL-Caltech)

Fig. 6.5 This is a close-up of a rock that has eroded from Whale Rock, showing the detail of the layers of deposition. There is some evidence of crossbedding in this rock, raising the question of whether it was deposited by water, like the lower elevations of the Pahrump Hills, or by dunes. (Image courtesy of NASA/JPL-Caltech)

Because this site was very important as the type site of the Pahrump member of the Murray formation, powder from the Mojave 2 sample was stored in two cups in the CHIMRA storage wheel. The second cup was held for several years before a newly modified SAM analysis was made, with dramatic results (reported in a later chapter).

The site Mojave (Fig. 6.6) was chosen for a sample. At the first place to drill (Mojave 1) the rock broke, so a nearby rock was drilled, producing the sample Mojave 2. The pale, flat rocks here and the sample were designated the Pahrump Hills member of the Murray formation.

Fig. 6.6 Curiosity took this selfie at the Pahrump Hills with its MAHLI camera. It was in the midst of the flat, exposed facies (geological term for face) of the fractured laminated slabs of rocks that were soon to become the type model of the Pahrump Hills member of the Murray formation. The selfie was photobombed by the peak of Mount Sharp, rising to the south. (Image courtesy of NASA/JPL-Caltech)

Fig. 6.7 Here is a view of Curiosity drilling sample Mojave 2 after the first Mojave try failed due to the breakage of a layer of rock. Many sols later, while drilling on the Vera Rubin Ridge, the topmost layer of rock shifted as the drill was raised. (Image courtesy of NASA/JPL-Caltech)

Fig. 6.8 This image was taken at Telegraph Peak, Pahrump Hills, by Curiosity's right front Hazard camera on sol 907. The mound on the left is Whale Rock. Curiosity is preparing to drill a core sample here. (Image courtesy of NASA/JPL-Caltech)

Telegraph Peak was the third and last contact science site chosen for Pahrump Hills. Analysis of the two earlier contact sites, each at lower elevations, had shown an unusual distribution of elements. The much higher Telegraph Peak site was requested for a comparison. See Fig. 6.8 for a view of the context of Telegraph Peak and its relation to Whale Rock.

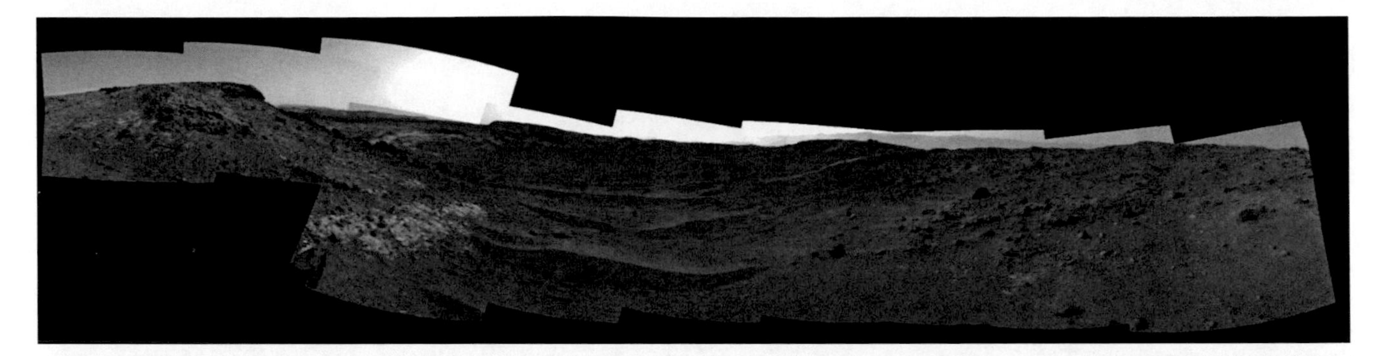

Fig. 6.9 Curiosity took the frames for this mosaic on sol 956, when it had left the Pahrump Hills and was halfway through Artist's Drive. It avoided the deep ripples of sand there by staying to the edge of the ripple field, crossing over when necessary. The elevation here is about 50 m above Bradbury Landing. (Image courtesy of NASA/JPL-Caltech)

Fig. 6.10 This shaded-relief topographic map shows Curiosity at sol 957 (green star) on Artist's Drive, about to bend left into a valley leading to Logan Pass. (Image courtesy of NASA/JPL-Caltech/MRO/MOL)

After taking the sample at Telegraph Peak, Curiosity was at the northeastern part of the Pahrump Hills area and proceeded to the south-southwest, examining several rocks. Curiosity then left Pahrump Hills along Artist's Drive (see Fig. 6.9). The track through the pass ran along the side of the deep sand ripples. On occasion, Curiosity crossed from one side of the valley to the other, moving in the micro-valley between ridges of sand. Near the end of Artist's Drive, the plan was to take a left fork into a valley that leads to Logan Pass. The goal was to explore a geological objective, a contact between two features (see Figs. 6.10 and 6.11).

Fig. 6.11 The sand in Logan Pass is obviously more intimidating than expected. Dark mountains and ridges rise above bright plates similar to those of the Pahrump Hills member of the Murray formation. At the top center of the image is Mount Stimson, named after Henry Stimson, U. S. Secretary of War during World War II. These dark features could be new members of the Murray formation or an entirely new formation. The contact between these two types of units is the current geologic goal. Since the direct route was impassable, the new plan was to back out of the valley that leads to Logan Pass and seek another route to the contact. Curiosity's right Navcam took this image on sol 984, May 14, 2015. (Image courtesy of NASA/JPL-Caltech)

Logan Pass was deemed to be too hazardous for Curiosity's mobility (see Fig. 6.11) Therefore, a new plan was formed to return to Artist's Drive, turn left, and take Marias Pass to seek the contact between the higher feature and the Murray formation there, along the higher feature's west slope. That nearly brings us to the next chapter of this tour, but first, the very important results of the campaign at the Pahrump Hills need to be discussed. Here are those results:

Results

The Murray Formation This was defined and named after reviewing the data collected in the three passes through the Pahrump Hills. The core sample at Mojave 2, with other observations, established the facies (the set of attributes that distinguish a formation or a member from others) of the Murray formation. The defining properties of Murray formation rocks are that they were formed in sediments of relatively still liquid water bodies such as lakes. The crystalline grains of the rock may be fine or coarse, but there should be a dominant component of the rock that is amorphous (particles are too small to show a diffraction pattern under X-rays). Such materials are termed mudstone. Such materials are also present in the Yellowknife formation on the floor of Gale crater, but they were formed in the flowing water of streams. The area of the Murray formation is limited to the north side of Mount Sharp because there is not enough surface evidence to extend it to other parts of Mount Sharp, at least until there is a further in-situ exploration mission.

Pahrump Hills Member The three samples of the Pahrump Hills area were similar enough in mineral composition, together with other observations, to tentatively declare a Pahrump Hills member of the Murray formation. However, the differences between the Pahrump Hills samples and the Buckskin sample were large enough to suggest establishing a new member. That will be discussed in the next chapter. Several members of the Murray formation have been defined in the course of this mission as Curiosity rose in elevation, distinguished by the facies of their bedrock plates. The first one, the Pahrump Hills member, is characterized by regular fractures, indicating shrinkage in the process of lithification (the sediment being made into rock). The flat surfaces of the rocks of the Pahrump Hills member are brighter than their broken edges, sometimes exposed in ridges. There could be a coating of evaporate or there could be bleaching by the combination of sunlight and solar wind.

Analysis of the Pahrump Hills member of the Murray formation samples was done by the SAM instrument suite at three sites. Two of them, at Confidence Hills and Mojave, showed evidence that some of the minerals were depleted, relative to others, specifically those that could be dissolved or that could be converted into other minerals, especially by slightly acidic water. The science team requested a third sample, at a higher elevation. The Telegraph Peak location (47 m above Bradbury Landing, 6 m above Mohave, and 7 m above Confidence Hills) was chosen. The result of that SAM analysis confirmed that the expected distribution of minerals was consistent with acidic water at about pH 4 being available to the lower elevation sites.

The conclusion was that the sediment at all three sites was similar but that acidic water came to the lower sites from groundwater before the sediment was compacted. This is an example of diagenesis, a process of changing the composition of sediment before or in the process of lithification. The process is particularly complex in the presence of mudstone. As it dries, pores are created between grains of other components, initially filled with water. When additional water permeates the material, it can contribute additional minerals to the pores, and they can interact with those already present.

The presence of acidic water is not necessarily toxic for life forms. In fact, some bacteria love it. In any case, this property of the Pahrump member of the Murray formation is a contribution to Goal I (life) and Goal III (geology) of the Mars Exploration Program.

CheMin Composition Results See Table 6.1 for the composition of the Pahrump Hill sites and the Buckskin site in the Marias Pass. Buckskin is the next higher sample site, discussed in the next chapter.

Table 6.1 Mineral and amorphous in weight percentage of samples in the Murray formation measured by CheMin

Mineral	Confidence Hills	Mojave 2	Telegraph Peak	Buckskin[a]
Plagioclase	20.4(2.3)	23.5(1.6)	27.1(2.8)	17.1(1.2)
Sanidine	5.0(0.7)	–	5.2(2.2)	3.4(0.7)
Forsterite	1.2(0.7)	0.2(0.8)	1.1(1.2)	–
Augite	6.4(2.2)	2.2(1.1)	–	–
Pigeonite	5.3(1.7)	4.6(0.7)	4.2(1.0)	–
Orthopyroxene	2.1(3.1)	–	3.4(2.6)	–
Magnetite	3.0(0.7)	3.0(0.6)	8.2(0.9)	2.8(0.3)
Hematite	6.8(1.5)	3.0(0.6)	1.1(0.5)	–
Quartz	0.7(0.5)	0.8(0.3)	0.9(0.4)	–
Cristobalite	–	–	7.3(1.7)	2.4(0.3)
Tridymite	–	–	–	13.6(0.8)
(H_3O^+, K^+, Na^+) Jarosite	1.1(0.7)	3.1(1.6)	1.5(1.8)	–
Anhydrite	–	–	–	0.7(0.2)
Fluorapatite	1.3(1.5)	1.8(1.0)	1.9(0.5)	–
Phyllosilicate[b]	7.6	4.7	–	–
Opal-CT	–	–	10.9	6.0
Amorphous	39.2 ± 15[b]	53[c]	27.2 ± 15[b]	54[c]

2-sigma errors are denoted in parentheses

Note a is the source of data for the Buckskin sample, Notes b and c are methods for the data on the amorphous component of the samples.

The composition of the three samples of the Pahrump Hills member and one additional sample of the Buckskin site in Marias Pass, the next higher sample on Mount Sharp, is to be discussed in the next chapter. *Left:* The column of strata includes formations that had not been defined when Curiosity was still approaching Marias Pass. (Source: E. B. Rampe et al., Mineralogy of an ancient lacustrine mudstone succession from the Murray formation, Gale crater, Mars, *Science Direct, Earth and Planetary Science Letters,* Volume 471, August 2017), pp. 172–185.

Shallow Gale Lake After considering all the evidence from the Pahrump Hills exploration, the hypothesis of a lake filling Gale Crater at least up to some level of the lower Mount Sharp gained strength as the working theory. Initially, it was thought of as a deep lake, but then the geologists realized that if sedimentation kept building up on the bottom of the lake, the depth could be modest while the surface of the lake continued to rise.

This revised understanding of Gale Lake was discussed after the Buckskin sample was analyzed, around sol 1000 in the summer of 2015. A very clear description of this new concept of a shallow Gale Lake at changing elevations was published in July, 2017 (E. B. Rampe et al., Mineralogy of an ancient lacustrine mudstone succession from the Murray formation, Gale crater, Mars, Science Direct, Earth and Planetary Science Letters, Volume 471, August 2017, pp. 172–185.) This paper was signed by 35 authors from 24 institutions. The paper included compositions from SAM results derived from Curiosity's sample from the Missoula contact at Marias Pass, Buckskin in Artist's Drive, and the sites at Pahrump Hills. The paper also includes CheMin observations of many other rocks. In this milestone paper, variations in minerology within different samples of the Murray formation were ascribed to modification (diagenesis) of the different members by washes of acidic water after the original mudstones were lithified.

As described in the legend of Fig. 7.13, successive floods can bring fresh sediment into Gale Lake in a series of steps, even covering Stimson sandstone with Murrray mudstone. Each flood would contribute a distinctive mix of minerals in the sediment it carried. Sources from the North would come from the Borealis Basin, caused by an ancient impact of a giant meteoroid or planetesimal, as proposed by Don Wilhelms and Steve Squyers in "The Martian hemispheric dichotomy may be due to a giant impact", *Nature,* Vol. 309, May 10, 1984. Gale Crater is on the very edge of the putative crater wall of the impact associated with the Borealis Basin. Immediately north of Gale Crater is the low-lying Aeolus Mensae, a series of fractures, possibly caused by the impact of the Gale crater event. They would be submerged in a Borealis Ocean if the north wall of Gale Crater was breached, as it must have been to form Peace Valley on that wall, the delta and fan associated with Peace Valley and, at its outflow, Yellowknife Bay.

Source of Floods To the South from Gale Crater are a number of radial V-shaped valleys leading directly to the rim of Gale Crater from a very large plateau called Terra Cimmeria on USGS Mars Topographic Map Aeolis, MC-23. A section of this map is shown in Fig. 1.5. The elevation at the source of the valleys is at +3000 m (relative to the average elevation of Mars).

In summary, there is clear evidence of water sources that could easily flood Gale Crater from all around its rim. The source of the water would come from a very large region with an area that is magnitudes larger than the area of Gale crater itself.

A late source of a flood, possibly the most recent, would be a low point in the north rim of the Gale Crater and a surge in Oceanus Borealis. The surge could have resulted from a tsunami caused by a meteor strike into the ocean that could send its water and sediment flowing to Peace Vallis to the delta and fan below (well-known from orbital photography) and ultimately to Yellowknife Bay.

Murray and Stimson Formations, Murray Buttes, and the Dune Campaign

7

After leaving the Pahrump Hills area, Curiosity attempted to proceed over Logan Pass and investigate the geology of the dark features above the Pahrump Hills member of the Murray formation at Mt. Stimson (see Fig. 6.11 of the last chapter). However, the Logan Pass appeared to be hazardous to Curiosity's mobility due to deep sand and loose rock at the edges. So Curiosity backed out of the path to Logan Pass and returned to Artist's Drive. The new plan was to continue on Artist's Drive to Marias Pass, (see Fig. 7.1) and then on to a new feature on the west side of the pass to investigate the contact between its base and another patch of bright outcrop, presumably the Pahrump Hills member.

Fig. 7.1 This panorama shows the Marias Pass area, selected for intensive investigation with a sample. This mosaic is composed of images taken by the Mastcam as Curiosity left Artist's Drive and entered Marias Pass. The dark, striated unit on the left is a foothill of the Apikuni Mountain. The similar bank on the right appears to be a similar formation, leading to Mount Shield (see the map in Fig. 7.2). After intensive exploration, the Marias Pass campaign, the area showed very high hydrogen and silica content, as well as a contact between the Pahrump Hills and the higher region to the left in this image. Curiosity's elevation here is 54 m above Bradbury Landing. Although it appears that the pale rocks are below Curiosity, they are really at the same level. (Image courtesy of NASA/JPL-Caltech/MSSS)

The Marias Pass Campaign

Curiosity drove 34 m to a site called Missoula (see Fig. 7.2) on sol 995. There, the dark younger feature (which could be either a new formation or a new member of the Murray formation) is in direct contact with the older Pahrump Hills member of the Murray formation. Curiosity would stay within a few meters

C. J. Byrne, *Travels with Curiosity*, https://doi.org/10.1007/978-3-030-53805-7_7

of this location until sol 1029. Remote science was performed for a few sols, including images of an eclipse of Mars satellite Phobos. Then there was a period when Earth and Mars were in conjunction, the sun between them, prohibiting secure communication, allowing only environmental observations.

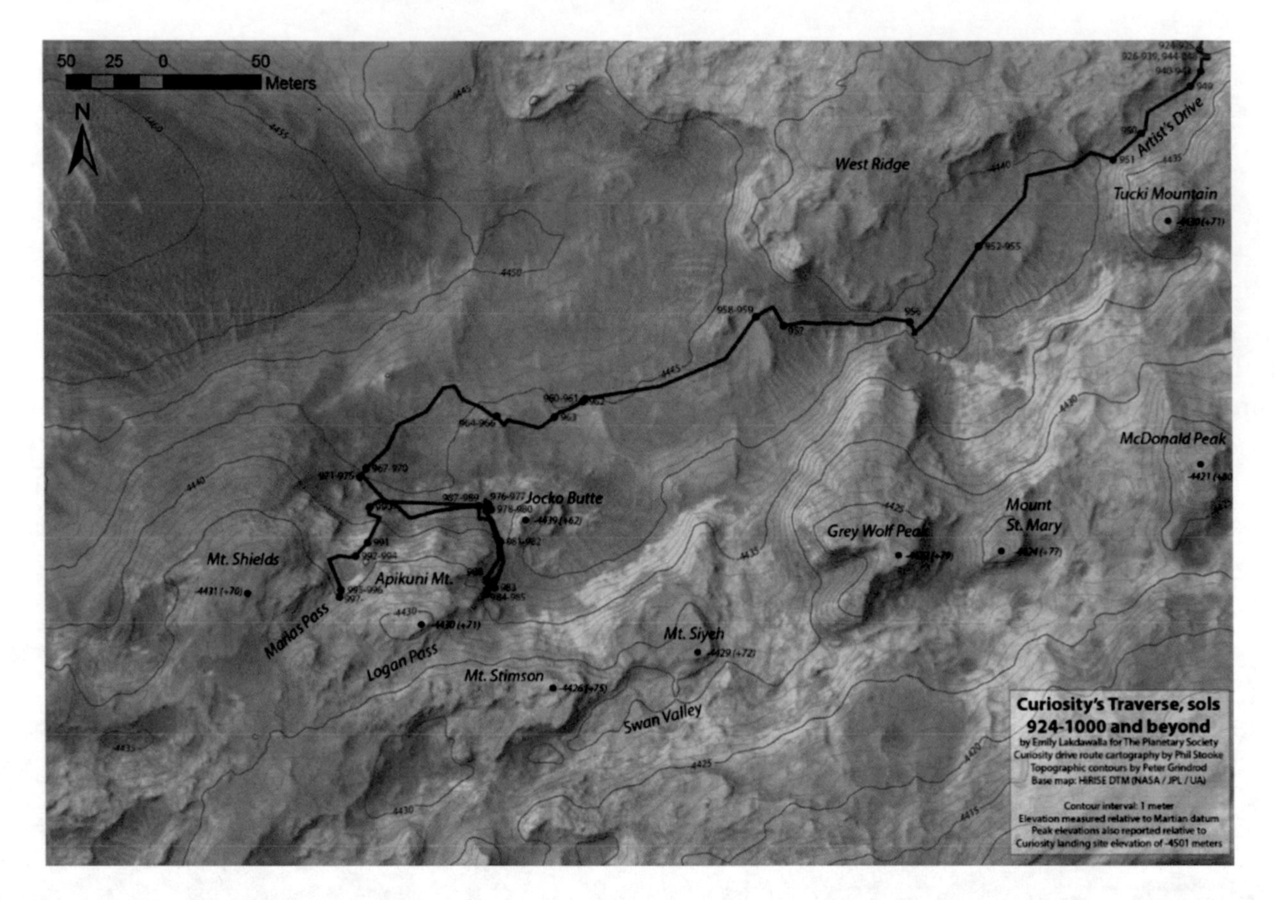

Fig. 7.2 Curiosity first tried to approach Mt. Stimson through Logan Pass to explore a new feature younger than the Pahrump Hills member of the Murray formation but was blocked by deep sand and rough rocks. It backtracked to Artist's Drive, turned left and entered Marias Pass to explore similar areas in contact with the Pahrump Hills member toward Mount/Shields (west) and Apikuni Mountain (east). (Image courtesy of: Planetary Society/Emily Lakdawalla/Phil Stooke/NASA/JPL-Caltech/HIRISE/UA)

After communication was restored, Curiosity took a drill sample at the Lion outcrop, near Missoula, where there is a contact between a foothill of Mount Shield and the Pahrump Hills member of the Murray formation (see Fig. 7.3).

As Curiosity drove to the Missoula contact, the DAN instrument had been continually operating in passive mode, monitoring neutrons received from the subsurface below Curiosity. Its neutron counts indicated a very high concentration of hydrogen atoms in the subsurface (see Fig. 7.3). Such a signal means that water molecules or hydroxyl ions are associated with minerals in the rocks, possibly clay or clay products such as the shale at the Shaler site.

Since there were two important factors present, the contact of a stratum above the Pahrump Hills member and the highest measurement of hydrogen yet by DAN, the science team decided to make the Marias Pass a major contact science area, including a drill sample. First, Curiosity retraced its path, for precise passive mode DAN measurements.

Altogether, Curiosity covered the Marias Pass area with three traverses, retracing its first path back to the northern entry to Marias Pass and including a new excursion to the east to examine an outcrop called

Fig. 7.3 Curiosity took this Mastcam image on sol 995, at the south end of Marias Pass after driving 34 m from its position at sol 992. This image shows a contact between the Pahrump Hills member of the Murray formation and a higher stratum (see text). Curiosity is still at the same elevation as it was when it took the image of Fig. 7.1, 54 m above Bradbury Landing. (Image courtesy of NASA/JPL-Caltech)

Lion. Further images were taken of interesting objects seen during the walk-about (see the image of the rock Lamoose in Fig. 7.4). The shape of this rock may be due to mud sediment deposited over a stream bed with shallow steps, such as eroded laminations of bed rock. The fractures could have happened as the mud dried. Lamoose was probably cemented by calcium sulfate carried by water from rain or drainage through sand from the early Stimson formation above and nearby, directly or through ground water. The diagonal scratches could have been made by wind-blown sand either before or after lithification (the process of being cemented into rock).

The single rock we call Lamoose is evidence of a series of alterations of repeated cycles of dry and wet climate in the history of this part of Mount Sharp, and by extension, the surface of Mars. The name is based on "bassin de la moose", a geological feature of the Moose River (Rivierre Moose on a French map) of Ontario, Canada. The Moose River flows into an arm of the Hudson Bay, near an island which where the Hudson Bay company established a trading post, the first English settlement in Ontario. The trading post served the fur trade, previously established by the French. The French translation for "la moose" is "le elan du Canada", that is, Canadian elk. Notice that "la moose" changes the gender of the word. The Lamoose rock is well named for it being evidence of change.

After confirming the DAN data and examining additional high silica rocks, Curiosity returned to the Missoula area for more high-resolution imagery (see Figs. 7.5 and 7.6).

Fig. 7.4 Lamoose rock, strangely shaped and abraded by wind, is very high in silica, like many rocks in the Marias Pass area. This one is about 10 cm across. The image is from sol 1041 near the very high hydrogen area. (Image courtesy of NASA/JPL-Caltech/MSSS)

Fig. 7.5 This is a closer view of the Missoula contact from Curiosity's Mastcam. The image shows the flat area similar to the Pahrump Hills member and darker rock above it. (Image courtesy of NASA/JPLCaltech/MSSS)

Fig. 7.6 Curiosity used the MAHLI hand lens for this mosaic of an end view of the Missoula contact zone. The image shows the edge of the flat area similar to the Pahrump Hills member and darker rock above it. (Image courtesy of NASA/JPL-Caltech/MSSS)

Fig. 7.7 Drilling for the Buckskin sample. The inset shows a mini drill hole to see how hard or soft the rock is. The result sets a parameter for the combination of rotation and percussion used in drilling. (Image courtesy of NASA/JPL-Caltech/KenKremer/kenkremer.com-Marco Di Lorenzo)

A fairly high elevation in the pale level rock of the Pahrump Hills member was selected for a drill sample at a site called Buckskin, near a foothill of the Apikuni Mountain called Lion. A picture of Curiosity's arm and turret as it deploys its drill is in Fig. 7.7. The resulting mini drill hole and tailings are shown in the inset. A mini drill hole is a test to see how hard or soft the rock is, to set up parameters for a full drill nearby.

After review of the contact images, and consideration of the high silica content and hydrogen concentration in the Marias Pass, there was a trend to considering the new unit a distinct formation, not a continuation of the Murray formation. Although this was not unanimous among geologists, the new formation was tentatively called the Stimson formation after nearby Mt. Stimson. Another example of the Stimson formation is the East wall of the Marias Pass, at the base of the West face of the Apikuni Mountain—a strongly cross-bedded sandstone that is shown in Fig. 7.8.

Fig. 7.8 Curiosity took the images for this mosaic of the Stimson formation from Marias Pass near the Buckskin sample site. It is strongly cross-bedded sandstone, possibly formed from fossilized dunes. The contact with paler rock, similar to the Pahrump Hills, is visible near the lower right corner of this image. (Image courtesy of NASA/JPL-Caltech/MSSS)

The Bagnold Dune Campaign

The striking results of the Marias Pass area, especially the images and sample results after taking the Buckskin sample and seeing the cross-bedding images of the Stimson formation, focused the science team on the issues of Martian dunes. From now on, the study of the dunes became at least as important as simply avoiding them for safe mobility. Curiosity proceeded on a long, scenic traverse through sandstone ridges of the Stimson formation toward the dune fields. Our journey will take us with Curiosity to visit Namib Dune and High Dune (see the map in Fig. 7.9)

Early in the Dunes campaign, Curiosity stopped at targets of opportunity called Big Sky (sol 1119) and Greenhorn (1137) to examine the chemistry of veins and vein halos in selected rocks (see Figs. 7.10 and 7.11). Fig. 7.11 shows all the drilll holes so far, including the two at Big Sky and Greenhorn.

Then Curiosity backed up to Meeteetse Overlook (sol 1144) for context images and some spectacular scenery (Fig. 7.12 and 7.13).

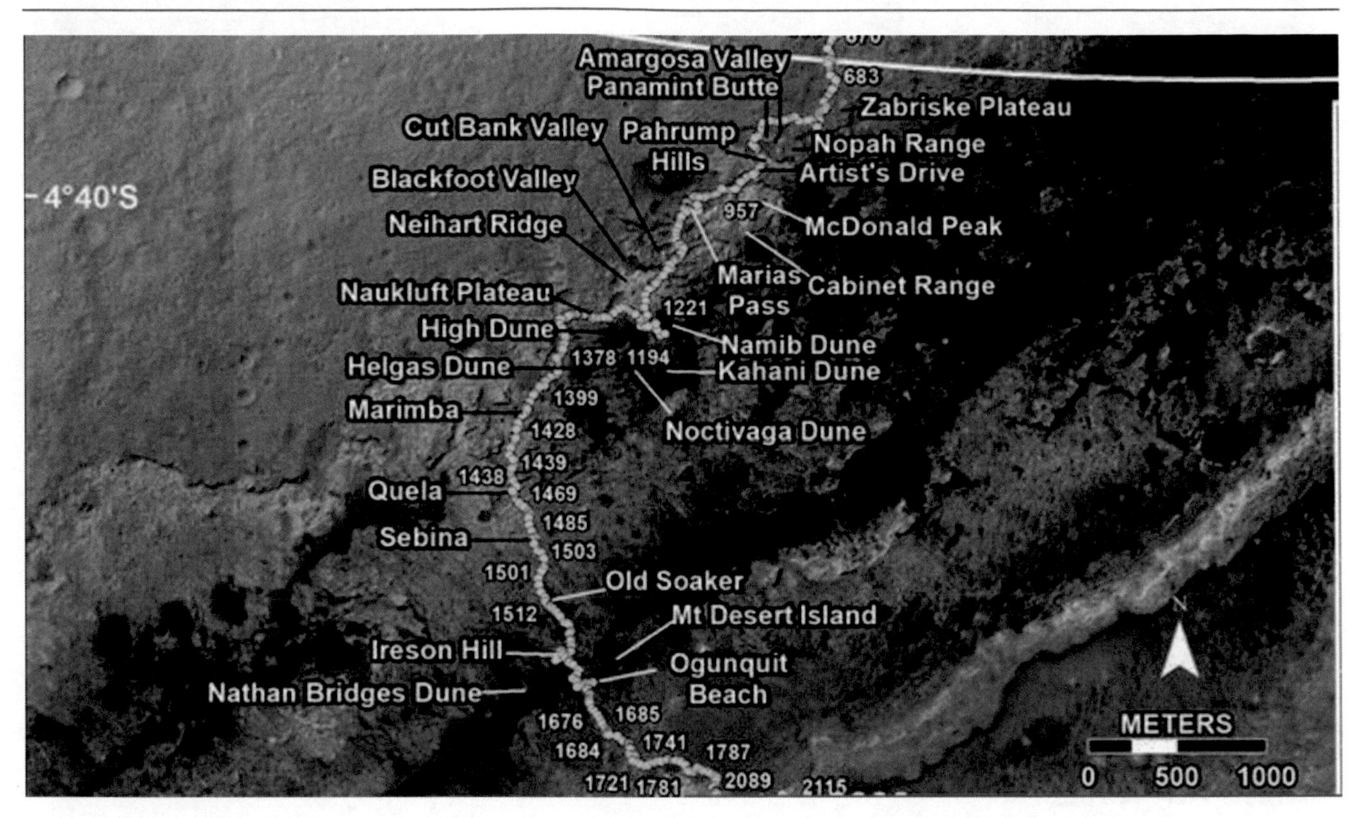

Fig. 7.9 Curiosity's path on lower Mount Sharp for the Dunes campaign. It passed by the areas of Namib Dune, High Dune, and the Murray Buttes (between the Marimba and Quela sample sites), continuing to Ogunquit Beach. (Image courtesy of NASA/JPL-Caltech)

Fig. 7.10 Mosaic of images showing the relation of the Greenhorn (left side) and Big Sky (right side) drill holes. Notice the light "halos" around the edges of the rocks on the left, probably due to intrusive seepage from veins associated with fractures in the rocks. Curiosity took the images on sol 1142. (Image courtesy of NASA/JPL-Caltech/MSSS/Emily Lakdawalla, from Planetary Society web page)

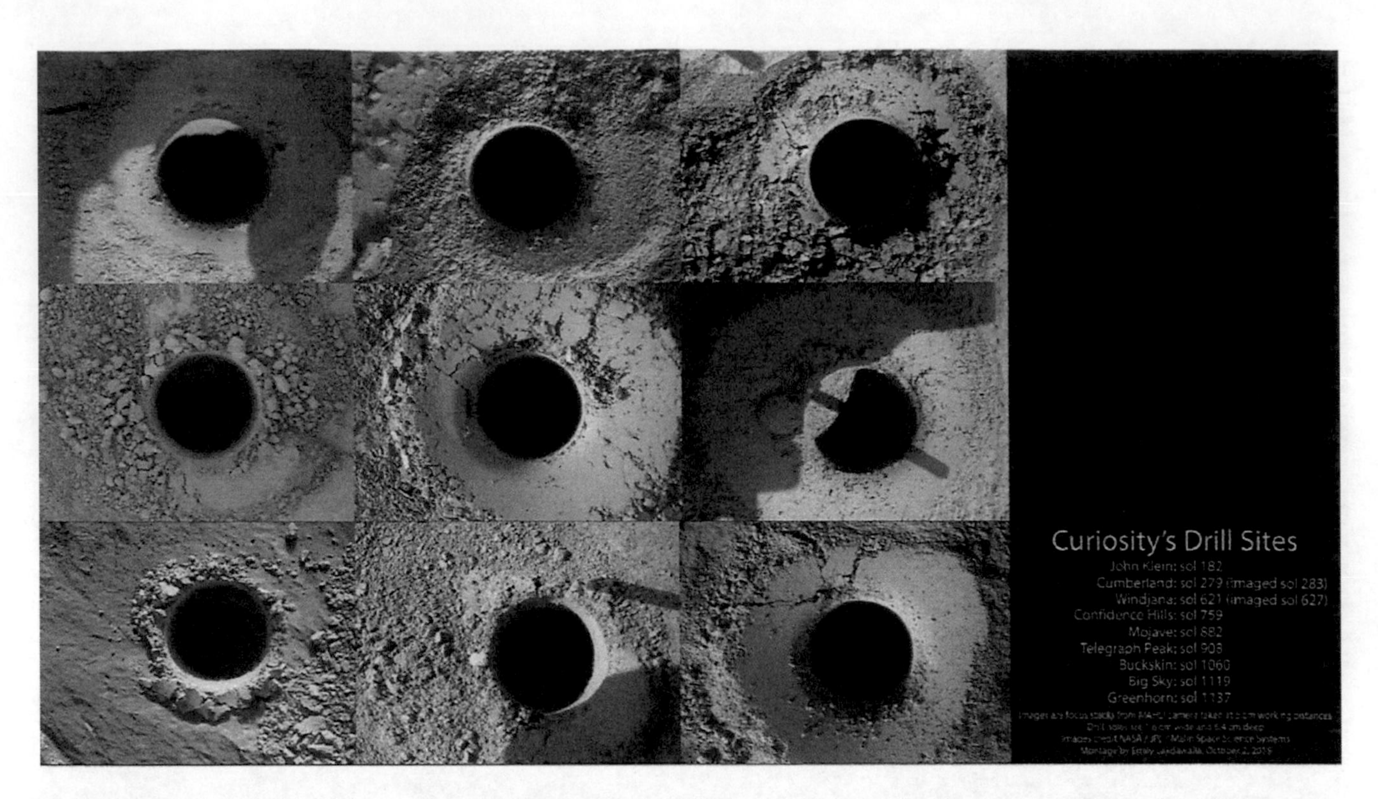

Fig. 7.11 All nine holes drilled so far are shown here. At the bottom center is Big Sky and the bottom right is Greenhorn. The two drill sites are within a meter of each other. (Image courtesy of NASA/JPL-Caltech/MSSS/Emily Lakdawalla, from Planetary Society web page)

Fig. 7.12 Meeteetse Overlook. This mosaic provides the context for the Stimson unit in the foreground, which rolls off to the right. Figure 7.13 provides a close-up of these rocks. The plates that cover the underlying unit turn smoothly downhill, as if they were cemented in place in the process of turning from something else into rock. The something else could have been a wind-driven sand dune. (Image courtesy of NASA/JPL-Caltech/MSSS)

Fig. 7.13 Blackfoot Valley and Cut Bank Valley, down from Mount Sharp to the west and off of our route. Curiosity is standing on the edge of the Stimson formation rocks seen from the Meeteese Overlook in Fig. 7.12, as they drop off to the valley below. The plates of rock here follow the steep slope down. This particular unit could be the downwind slope of a wind-driven sandy dune. The frames of this mosaic were taken by Curiosity on sol 1148. The elevation here is 67 m abosve Bradbury Landing, 13 m higher than the Buckskin site. (Image courtesy of NASA/JPLCaltech/MSSS)

Curiosity took the frames of this pair of spectacular mosaics of Meeteetse Overlook, Blackfoot Valley and Cut Bank Valley in order to document the nature of the Stimson formation. The mosaic in Fig. 7.12 was taken on sol 1144 at Meeteetse Overlook looking south. It shows the Neihart Ridge in the background, with the edge of a Stimson feature at the left, dropping down to the west. Curiosity took the image of Fig. 7.13 from that same Stimson area, its front wheels at the edge of the drop-off and the camera looking west. Figure 7.13 shows Blackfoot Valley, leading to Cut Bank Valley.

The names chosen for features in the Marias Pass are related. The Blackfoot Valley on Earth, in Montana, USA, was virtually a character in the motion picture "A River Runs Through It", directed and produced by Robert Redford. The movie was based on the book by Norman Maclean, who grew up there. The Blackfoot River ends near the city of Missoula, Montana. On Mars, Missoula is the name of a contact between the Murray and Stimson formations in Marias Pass, near Curiosity's position when she took the picture of Fig. 7.13. The source of the Blackfoot River is near Meeteetse, Wyoming, on the border of Montana and Wyoming. The Lewis and Clark trail passed there.

The Cut Bank Valley, also in Montana, is in Glacier National Park, where there is a mountain named for Henry Stimson, for whom the Stimson formation is named. He helped survey the park and was a supporter of its establishment. Much later, he was The U. S. Secretary of State in World War II.

The view of these valleys is spectacular, but we are not going to follow them off to the west; our path continues to the southwest. In ancient times, these valleys probably carried water through a system of valleys that Curiosity had been following since she left Piute, near the Armagosa Valley, before the Pahrump Hills campaign. She left those valleys for a higher elevation at sol 1073. The next stop will be near High Dune, where the objective was to study how these dunes relate to the ancient ones that became cross-bedded ridges.

The Dune Campaign: Phase 1

The Dune campaign is a series of studies of active Martian dunes that we will encounter and pass on our way south and upward on Mount Sharp. The science team had major motivations for this study. They had evidence on the floor of Gale crater and the lower Mount Sharp that rocks of the Murray formation had begun as deposits in still water—mudstones that started as sediments in a lake. However, there were ridges with cross-bedded structures that appeared to be dunes of wind-driven sand, laid down in a dry environment.

In order to resolve these conflicting ideas of a deep lake and a waterless wind, it seemed to be a good idea to get as much information as possible on what active dunes in Mars were like. What are their shapes? What is the sand like in composition and grain size? Where did it come from? How were the dunes turned to rock? Could cross-bedded rocks be formed in water some way?

A second motivation was the basic explorer's instinct. Here is an opportunity to be the first to intensely study active dunes in situ on a planet other than Earth. Wow! So the Bagnold Dunes changed roles from being an obstacle to overcome to being an important subject of study.

There are two types of dunes studied here. First, Curiosity came to barchan dunes (Phase 1) and then to linear dunes (Phase 2). Examples will be given of each type as we come to them. See the map in Fig. 7.14 for the positions of dunes of each phase of the campaign. Note also that the Bagnold dune field is split into northern and southern branches.

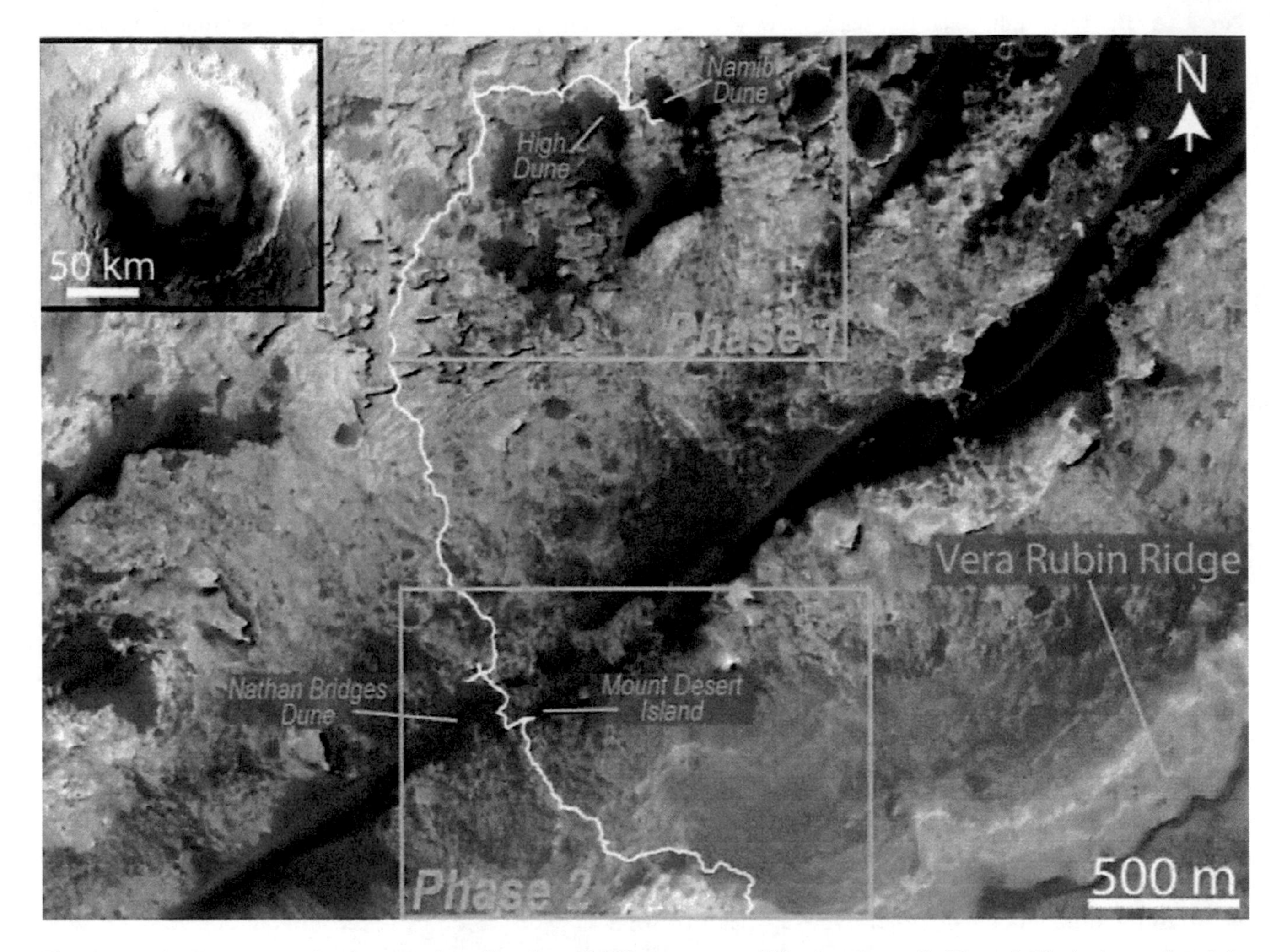

Fig. 7.14 The Dune campaign was a series of studies of different types of Martian dunes. In Phase 1, Curiosity explored the nature of barchan dunes and then traveled past the Naukluft Plateau to pass between fields of linear dunes in Phase 2. This image, with Curiosity's track from JPL-Caltech, was taken from the open access paper of M. G. A. Lapotre and E. B. Rampe, "Curiosity's investigation of the Bagnold dunes, Gale crater overview of the two-phase scientific campaign and introduction to the special collection", Geophysical Research Letters 45, Issue 19, October 16, 2018.The base image is from the HRSC camera of the European Space Agency's orbital Mars Express spacecraft, processed by the German Aerospace Center (DLR). (Image courtesy of NASA/JPL-Caltech/ESA/HRSC/DLR/GRL (open access))

Curiosity first examined High Dune (Fig. 7.15) and then took a side trip to closely explore the downwind side of Namib Dune, another active barchan dune (Figs. 7.16, 7.17, and 7.18). It returned to High Dune and went westward and then southward along the eastern edge of the Naukluft Plateau by way of the Murray Buttes (Fig. 7.19) to get around the end of the northern branch of the Bagnold dune field. Curiosity then left the Phase 1 area and drove to toward Phase II.

Namib Dune is the next stop on our route. Curiosity's path went southwest from the Meeteetse Overlook and took a short run to the east to approach the downwind end of the dune, a total of 55 m. Figure 7.16 shows the Namib Dune from the north, with the peak of Mount Sharp photobombing the selfie.

Figure 7.17 shows a close view of this dune, with the characteristic features of barchan dunes labeled.

Fig. 7.15 High Dune is just a short distance west of Namib Dune in the northern branch at the Bagnold Dune field. After taking this picture of High Dune on sol 1115, Curiosity took a side trip to explore Namib Dune (see Figs. 7.16, 7.17, and 7.18). Curiosity then returned here. Its path continued west and then south, to go around High Dune. It then passed along the eastern edge of the Naukluft Plateau. Part of that plateau can be seen in this image near the upper right corner of the image and behind the dune. (Image courtesy of NASA/JPL-Caltech/MSSS)

Fig. 7.16 Namib Dune is shown in this 360 degree panorama Curiosity took with its Mastcam on sol 1157. The camera was about 7 m from the base of the dune. The dune is about 5 m high. (Image courtesy of NASA/JPL-Caltech/MSSS)

Fig. 7.17 Namib Dune is a barchan dune, with its typical characteristics shown. (Image courtesy of NASA/JPL-Caltech/Planetary Society/Emily Lakdawalla)

Fig. 7.18 Curiosity took this image of small ripples on top of large ripples on top of Namib Dune. (Image courtesy of NASA/JPL-Caltech/MSSS)

Many of the barchans and linear dune fields exhibit small ripples superimposed on larger ripples, in both crest-to-trough and wavelength measures (see an example in Fig. 7.18). These parameters are discrete, not evenly distributed in a spectrum. This characteristic is common on Mars, but unknown on Earth. The reasons for this difference are for further study.

Murray Buttes On Earth or Mars, buttes are formed by strata of relatively hard material over strata of softer, more easily eroded material. Erosion can result from flowing water or wind. Freeze and thaw cycles are effective to increase the rate of erosion. As the eroded material crumbles, the hard material is undercut until it, too, falls away. What is left is often flat on top, with a nearly vertical boundary falling to a skirt of loose material (see the butte on the right side of Fig. 7.19). This debris from erosion lies at an angle that depends on gravity and the friction of the material, as well as grain size, and wind.

This loose, slanted rock is called a talus slope in geology and a scree slope in hiking. In either case, it is difficult to traverse; a small disturbance can cause an avalanche. Therefore, exploration of buttes is difficult. So as interesting as it may be to climb up a talus slope to find out why the top is so different than the underlying strata, Curiosity has not driven to the top of a butte.

Fig. 7.19 Here, at waystop 1387, Curiosity rejoined the Lewis and Clark Trail (path planned in September 2014) after leaving it on sol 1281 to move west to visit the Naukluft Plateau. After sol 1387, Curiosity stayed on the Lewis and Clark Trail through the Murray Buttes area, moving south. The scarp that runs across this image was broken down by a pair of meteorite craters, possibly secondaries from a larger impact. The ridge in the foreground is the rim of one of the craters. The other crater is beyond the break in the scarp. Note the balanced rock on the point in the left background. This image was taken by Curiosity's Mastcam. (Image courtesy of NASA/JPL-Caltech/MSSS)

Fig. 7.20 Curiosity waved a greeting with her arm as she took the images for this mosaic with her Mastcam. The rover is about to enter the Murray Buttes area. This mosaic is part of a panorama whose images were taken around sol 1405, Earth date, July 17, 2016. (Image courtesy of NASA/JPL-Caltech/MSSS)

The Murray Buttes are examples of the Stimson Formation: This area had been a major goal of the MSL Science Team since Bradbury Landing and exploration of Yellowknife Bay. Ironically, the exploration of the Murray Formation at Pahrump Hills and the Stimson Formation near Marias Pass was successfully begun before entering the Murray Buttes. It was understood by then that the buttes were examples of the Stimson formation, mostly sandstone that had been deposited by windblown sand.

It is clear from Fig. 7.20 that passing through the Murray Buttes will remind us of the appearance of much of the American Southwest. The common element is deep strata of sandstone, deposited as sediment in generally horizontal strata, that subsequently erodes in an arid climate.

To understand the processes at work, it would be helpful to view the Murray Buttes from above. Now how can we do that? Oh, I know, we have spacecraft in orbit around Mars! Figure 7.21 is a shaded relief map of the Murray Butte area. The Murray Buttes, like the Murray formation, are named after Professor Bruce Murray, a Caltech planetary geologist who created a geologic map of Mars and was the Director of JPL between 1976 and 1982

Butte or Mesa? Seen from a plain, a butte or a mesa looks the same as a scarp of a plateau. All three forms have primarily flat tops; the major differences are whether the unit is isolated (butte or mesa) or the edge of a plateau (scarp). In time, erosion may cause the edge of a plateau to retreat, leaving a field of mesas or buttes where the cap rock is a little harder than the surrounding strata. The distinction between a butte and a mesa is not clearly definable. A butte is said to be smaller than a mesa and sometimes said to be narrower than a mesa. The origin of the distinction in terms is said to be in the American southwest; see box. Of course, that definition is not useful on Mars. Our restricted definition for this book is "If it is in the area of the Murray Buttes, it is a butte, not a mesa". Elsewhere on Mount Sharp it can be either.

Butte or Mesa?

(pragmatic distinction from U. S. southwest)
"If you can herd cattle on it, it's a mesa."

Looking at the shaded relier map of Fig. 7.21, we can see that there was once a layer of easily erodible material over a more resistant layer (the same Murray Formation Curiosity sampled at Pahrump Hills). The original eroded layer of sediment was about 15 m (40 feet) high.

Fig. 7.21 Murray Buttes Map. This map of the Murray Buttes area is based on a photograph from orbit, with the slopes emphasized by shaded relief, as if the sun were shining from the Northwest. The Mars Reconnaissance Orbiter has not only a high-resolution camera, but also a laser altimeter. The brightness of the map is albedo, not photographic brightness, and the shaded relief is generated from a topographic model of the surface. (Image courtesy of NASA/JPL-Caltech/LRO)

Caprock is shown in close-up profile in Fig. 7.22. This is one of the images used in the panorama mosaic of Fig. 7.20. The close-up of the caprock of Fig. 7.22 is at the top left of the butte to the right of the turret on Curiosity's arm in Fig. 7.20. Buttes are capped with unusually hard rock that protects the rock below from being eroded. It may have originally been of the same sandy material as the sediment below it but then been flooded with water saturated with a cementing material that filled-in pores in the sediment, bonding chemically and cementing the silicon oxide and other crystals as the water flowed off and evaporated. Most likely the cementing substance at lower Mount Sharp was largely calcium sulfate.

The talus slope extending across the picture of Fig. 7.23 from the left edge of the image is from a much larger butte; part of its caprock is shown in Fig. 7.24.

Fig. 7.22 This is the caprock at the left end of the butte on the right of Curiosity's turret in the panorama in Fig. 7.20. Note that the hardening of the rock was uneven, leaving thin laminations. (Image courtesy of NASA/JPL-Caltech)

Fig. 7.23 This is a much smaller butte than the one whose caprock was imaged in Fig. 7.22. This image is from a single frame of Curiosity's Mastcam when she was about 30 m away from the butte, on sol 1432, having driven about 300 m from where she took the frames for the panorama of Fig. 7.20. Although small, it has the typical structure of a butte, flat top, caprock, nearly vertical eroded wall and a talus slope of accumulated loose eroded debris. (Image courtesy of NASA/JPL-Caltech)

Fig. 7.24 End caprock of the larger butte next to the small butte shown in Fig. 7.23. (Image courtesy of NASA/JPL-Caltech). Notice the patch of sky that can be seen between a boulder and slabs of fallen rocks above the top of the talus slope. (Image courtesy of NASA/JPL-Caltech)

After driving 55 m to the south on sol 1433, Curiosity took images with her Mastcam back towards her last waystop (see Figs. 7.25 and 7.26). The small mosaic in Fig. 7.25 included the valley floor she had been following and the pair of buttes, small and large, west of the valley floor as well as a very large butte east of the valley floor. This butte had two levels of caprock, a low one and a higher one. The geologic term for the result is called a bench.

Fig. 7.25 A look back north from the next waystop after sol 1432, on sol 1433. This is the valley among buttes that Curiosity had passed through the previous sol. The two frames of this mosaic were taken by the Mastcam from about 55 m (180 feet), the length of the last drive. The gray talus slope on the left side of the mosaic is from the larger butte, the neighbor of the smaller butte seen in full in Fig. 7.23. (Image courtesy of NASA/JPL-Caltech, mosaic by Charles Byrne)

In the left foreground, there is the gray talus slope of a medium butte. Near the left top corner of the image, slightly redder caprock of a small butte appears, close behind the talus slope. This is the butte that was seen in full in Fig. 7.23, with the talus slope of the medium butte in the foreground. Curiosity took another picture looking north, back to her last stop. Figure 7.26 is a second picture of the caprock of the larger butte seen from the east in Fig. 7.24 of the end caprock of the larger butte (see Fig. 7.26).

Images for another mosaic, this one taken toward the south-southeast and covering nearly 90° (Fig. 7.27) were taken on sol 1434. Note the peak of Mount Sharp photo-bombing again between two buttes.

The butte on the right of Fig. 7.27 is 70 m long and thin, less than 10 m wide. How it came about is an interesting question.

The next mosaic was taken between sols 1448 anlld 1451. It shows the trail Curiosity will take as it almost leaves the Murray Buttes area, but it still has important work to do there.

Butte Close-up Curiosity took a picture of the left butte in the mosaic of Fig. 7.28 on sol 1428 with her Navcam (see Fig. 7.29). It is taken from a slightly different angle than the Mastcam images.

After going through the pass between the two buttes shown in Fig. 7.28, Curiosity stopped, did a U-turn, went back a short disance, and turned west, going back behind the butte on the right of the mosaic in Fig. 7.28. There, she celebrated the completion of the Murray Buttes campaign by taking a selfie (Fig. 7.30). For good measure, she took a drill sample as a souvenier (National Parks regulations do not apply).

Fig. 7.26 This is the same end caprock seen in Fig. 7.24, which also showed a much smaller window between the detached bolder and the rest of the endcap. This picture was taken from the south on sol 1433 with the Mastcam and Fig. 7.24 from the east, about 90° away, so most of the bolder was hidden. Initially I thought that this was just a boulder, but as I stared at it, I thought I saw the profile of a Martian, wearing a hat. Do you see that, too? (Image courtesy of NASA/JPL-Caltech)

Fig. 7.27 Mosaic of nearly 90°, east-southeast from waystops 1434. The two closest buttes are about 90 meters away, the others are much further. Curiosity's location is about the middle of the Murray Buttes area. (Image courtesy of NASA/JPL-Caltech/MSSS).cap

Fig. 7.28 Curiosity took the images for this mosaic with her Mastcam, in the southern part of the Murray Buttes. After she took the images for this mosaic, her path ran through the pass between the two buttes, towards the one in the distance. (Image courtesy of NASA/JPL-Caltech/MSSS)

Fig. 7.29 shows a butte whose caprock has degraded but is still doing its job protecting the much softer rock below. (Image courtesy of NASA/JPL-Caltech/MSSS)

Fig. 7.30 Curiosity took this selfie at the waystop and drill site called Quela. She is saying "Goodbye" to the Murray Buttes, driving south for Phase 2 of the Bagnold Dunes Campaign. The drill sample name Quela, like other names in the Murray Buttes, was chosen from the map of Angola. She waved her arm in greeting to the Murray Buttes area on sol 1405 and left it on sol 1464. (Image courtesy of NASA/JPL-Caltech)

Dune Campaign Phase 2

Old Soaker As Curiosity continued upward beyond the Murray Buttes, a target of opportunity appeared. As can be seen in Fig. 7.31, the patch of Old Soaker rocks resembles a patch of mud left by a drying pond. As mud dries, losing water to the atmosphere, its surface contracts until polygonal cracks appear. The surface, having lost water, is heavier than the underlying wet material, forcing it to ooze up into the cracks. That is what we see in Fig. 7.31. It remains to wonder what settled on top of this hardened surface to preserve the pattern, just as fossils are preserved in Earth rocks.

After studying Old Soaker, Curiosity proceeded southeast to cross the linear dunes of the southern branch of the Bagnold dune field. The path starts between the Nathan Bridges Dune (Fig. 7.32) and The Enchanted Island dune field (Fig. 7.33). Dr. Nathan Bridges led the planning of the dune campaign. He died unexpectedly before it was completed.

Fig. 7.31 Curiosity took this image of Old Soaker with her Mastcam on sol 1555. The rock is 80 cm across and shows patterns of cracks that are similar to those of mud on Earth shrinking as it dries. The cracks are filled by sediments from below and become ridges. (Image courtesy of NASA/JPLCaltech)

The first dune field visited in Phase II of the Bagnol Dune Campaign was named after Dr. Nathan Bridges, who was a leader in planning the campign who had studied dunes on Earth, on Mars, from the surface and from orbit. He died, unexpectedly, on April 27, 2017.

This linear dune has regular, long wavelength ripples and low short wavelength ripples at right angles.

Curiosity scuffed some of the ripples in the sand with its wheel to expose a cross section of the ripple.

The image at Ogonquit Beach (Fig. 7.34) showed the last of the linear dunes of Phase II of the Bagnold Dunes campaign. A scoop sample was taken of the sand so SAM could examine the composition of a linear dune.

Fig. 7.32 Nathan Bridges Dune is linear with large and small ripples. Dr. Bridges (1966-2017) was a leader of the Bagnold Dune Campaign. Curiosity took the picture from the north, with Mount Sharp in the background. This is a mosaic of frames from Mastcam. (Image courtesy of NASA/JPLCaltech/MSSS)

Fig. 7.33 Enchanted Island dune field shows large ripples transverse to wind directions but also small ripples at right angles to the larger ones. Mastcam mosaic image was taken on sol 1752. (Image courtesy of NASA/JPL-Caltech/MSSS)

Fig. 7.34 The path continues to the Ogonquit Beach ripple field, where contact science was performed, including a scoop sample of the sand for SAM analysis. (Image courtesy of NASA/JPL-Caltech/MSSS)

Results

The Bagnold dune campaign lasted from November 2015 to April 2017. Of course, other science was also conducted, but the study of the interaction between sand and wind took precedence in this period. Once Curiosity arrived at Ogunquit Beach, the Bagnold Dune Field was no longer in the way of moving directly upward on Mount Sharp. At Ogunquit Beach, Curiosity was at an elevation 201 m above Bradbury Landing on March 9, 2017, 1,648 sols after landing, and had traveled 16,128 m. The first thorough examination of such dune properties as grain size, waves, and ripples of active dunes on Mars had been accomplished.

Strata Higher Than Murray Observed In the time before the Buckskin drill sample taken at Marias Pass, at the beginning of this chapter, the growing consensus was that the Murray formation of lake-deposited mudstone was the dominant form of rock on lower Mt. Sharp, perhaps all the way up to the Vera Rubin Ridge and maybe the clay area (confirmed later). In this section the results from Buckskin to the end of the Bagnold dunes campaign will be summarized. In the next chapter a new paradigm for the strata relationships accepted by the MSL Science Team in 2018 will be described. That new way of understanding the results of this chapter is a major accomplishment of the science team. In a later chapter, the tour along Curiosity's path will be resumed.

Cross-Bedding As seen in many ridges along Curiosity's path (see Fig. 7.8), cross-bedding was thought to be associated with dunes that had been turned into rocks. Some geologists claim the dunes were made of coarse grained sand, transported by wind (alluvial), in dry environments.

Other geologists were aware of rock that had been deposited in slow flowing water (fluvial) and were cross-bedded, the flowing water having played the role of wind. In this case, the typical grain size is smaller, like silt.

At this point, the two groups of geologists started to diverge, one group favoring dunes having been formed in a dry environment and the other favoring deposition of fine-grained sediment and silt. A positive result of the competition was the definition of the Bagnold dunes campaign, increasing the priority of the need for collecting data about both currently active dunes and crossbedded rock units to relate them to environmental influences. While the dunes campaign was being carried out, these two contrasting views were presented in the abstracts and talks of the 2017 Lunar and Planetary Science Conference and peer-reviewed papers.

Buckskin Drill Sample Results of the SAM analysis of the Buckskin sample were compared with samples from the Pahrump Hills area. The main differences are that Buckskin's larger mudstone component is richer in water and silica than previous mudstone samples analyzed by SAM. This finding was supported by other instruments of Curiosity. Silica is silicon dioxide, common in sand. In pure form it is called quartz. It is weakly soluble in water.

Marias Pass Member of the Murray Formation Because of the differences in composition with the Pahrump Hills member of the Murray formation, the term Marias Pass member was assigned to the stratum.

Missoula Contact Above the flat layer of Marias Pass member rock that was sampled at Buckskin, there was a different stratum of rock that was darker, and that made an irregular contact with the Murray formation rock. It was called the Missoula contact (see Figs. 7.5 and 7.6). Because of the irregularity in the Missoula contact, it has the character of an anomaly; that is, there may have been erosion or a change of environmental conditions between the Stimson formation and the underlying Marias Pass member of the Murray formation. That implies an uncertainty in the duration of time between the formation of the units on either side of the contacts and uncertainty of changes in the environment. Further observations have confirmed the data collected at Marias Pass.

Stimson Formation Across the pass from the Buckskin site was a tall ridge of dark rock whose sediment layers were cross-bedded (see Fig. 7.8), unlike the Pahrump Hills member of the Murray formation. Like the stratum above the Marias Pass member, both units were designated the label Stimson formation, named after Henry L. Stimson, who helped establish Glacier National Park. The ridge of Fig. 7.8 is part of Apikuni Mountain, which is across the Logan Pass from Mt. Stimson. The rocks of the Stimson formation at Marias Pass are sandstone, of a coarse grain.

Big Sky and Greenhorn Samples As Curiosity climbed further along and higher beyond the Marias Pass, fractures in the flat bedrock of the Marias Hills member began to show halos around the fracture, whitish coatings on the plate edges alongside the fractures. These halos were evidence of modifications of the rock and were hoped to provide information on the high level of silica in previous samples.

Two samples were taken from drill holes about a meter apart, Greenhorn in a halo and Big Sky with no halo, to see if analysis would lead to new information. Comparison of the two distributions of molecular abundances showed that subtracting calcium oxide and silicon oxide (silica) from Greenhorn (the one from the halo) would improve the fit. Although there are alternate ideas, one possible conclusion is that the excess silica in Greenhorn is modification due to groundwater seeping through the fractures in the rocks.

Old Soaker The appearance of this rock (Fig. 7.31) indicates that a mud puddle dried up with the surface having a pattern of fractures, separation of the surface skin, and material oozing up along the fractures. The pattern is similar to observations on Earth and indicates variable wetting and drying cycles before deposit and subsequent erosion of a covering layer.

Bagdold Dune Campaign This was the first systematic investigation of linear and barchan dunes in a planet other than Earth. Similarities and differences were found in the climate of Mars, the movement of dunes, primary and secondary ripples,and changes in wind direction. Grain size distributions were found in diverse dune fields.

8 Two Formations: Deposited and Eroded

After definition of the Murray formation as a mudstone deposited in water, based on the exploration of the Pahrump Hills area, there was speculation that the rock further up Mount Sharp would be similar. Lake Gale was envisioned as a deep lake, with some sediment deposited on the floor of Gale crater and some on the shore of Mount Sharp. The lake was assumed to rise.

The nature of Lake Gale was modified at least twice as Curiosity explored more areas. At Kimberly, the discovery of manganese oxide indicated that the environment was oxidizing. However, at both Pahrump Hills and Big Sky there was evidence that a reducing environment had modified the abundance of certain metals. In response to this new evidence, the suggested model of Lake Gale was modified to allow for different parts of the lake to have different characteristics (see Fig. 8.1).

Shallow Gale Lake? It is also possible that Gale Lake was always a shallow lake over a bed that rose with the sediment of each flood event (see the Result section of Chap. 6). This model would produce a lake surface (and shore elevation) that would have left the series of outcrops of members of the Murray formation that are observed. The composition of each deposit would change with the nature of each flood. A series of shallow Gale Lakes, rising in elevation as the bed rises, would also facilitate moving sediment from the rim of Gale crater toward Mount Sharp. The action of wind-driven waves would distribute the sediment, causing some sediment to be moved to a shore at Mount Sharp. This is similar to the process of moving sand from an ocean to a beach.

As Curiosity explored further, especially during the dune campaign, the situation became more complex. The geologists in the science team split into two groups with different views of the cross-bedded units. One group believed they were all derived from wind-driven coarse-grained sand dunes in an arid climate. The other believed that at least some cross-bedded areas were very fine-grained sand or silt deposited in flowing water. The first group would assign all cross-bedded sections to the Stimson formation. The second would keep some cross-bedded units in the Murray formation. This situation was clear in the abstracts for the 2017 Lunar and Planetary Science Conference. Those abstracts were submitted in January 2017 and the talks given in March 2017.

As a result of the contributions exchanged at the conference the different views were reconciled. A new group of formations was proposed: the Siccar Point group, based on concepts originated by the geology of the Siccar Point rocks on the western coast of Scotland. The Mars rock sections that were derived from coarse and medium grained sandstone units, cross-grained or not, were assigned to the Stimson formation. Mudstone rocks, cross-grained or not, were assigned to the Murray formation. Note that most rocks have both sand (crystal grains that show an X-ray diffraction pattern) and silt (amorphous particles that do not show an X-ray pattern); the difference between the designations mudstone and sandstone depends on which component dominates.

C. J. Byrne, *Travels with Curiosity*, https://doi.org/10.1007/978-3-030-53805-7_8

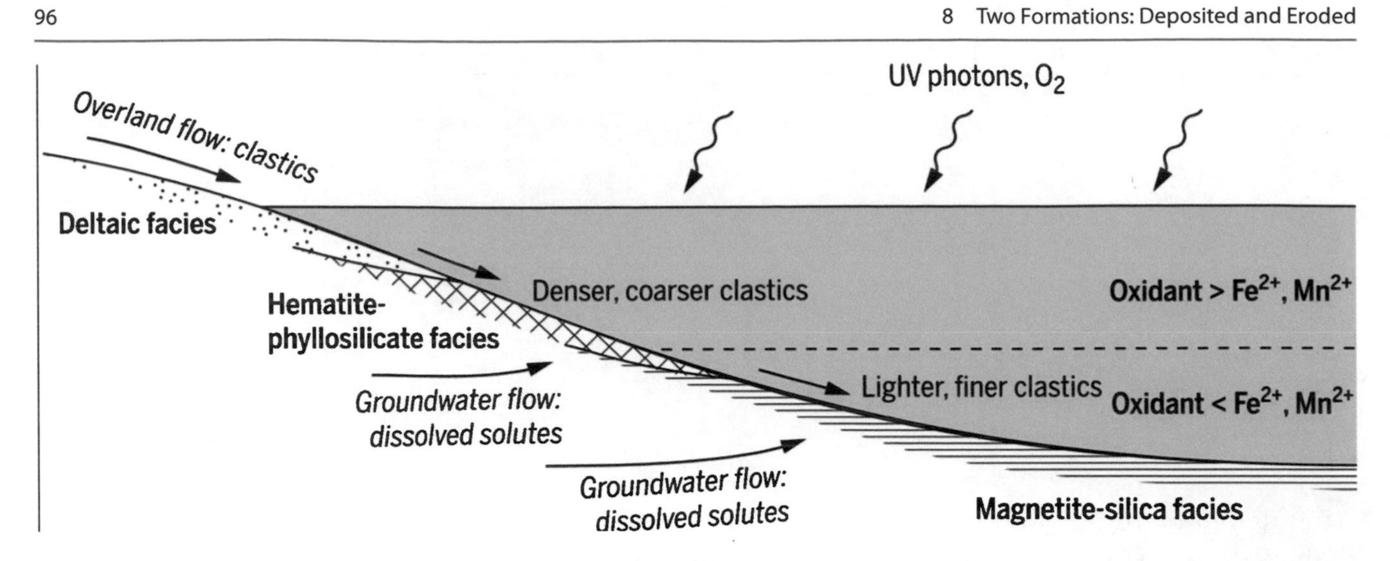

Fig. 8.1 An early view of Gale Lake, based on Curiosity data from Yellowknife Bay and the Kimberly area. This image is from a Research Article published in Science in June, 2017.The lead author was Joel Hurowitz of Stony Brook University. It shows a deep lake in Gale crater, with a neutral Ph. Oxygen entered the surface of the water and was combined with minerals brought in by floods and groundwater. As Curiosity gathered more data at Kimberly, Pahrump Hills and Big Sky, it was suggested that an oxidizing environment could be found near the surface and a reducing environment at depth. (Image courtesy of NASA/JPL-Caltech/Stony Brook University)

The climate changed to arid as time went on, perhaps near the end of the Noachian Period, and the Gale Lake level receded. The part of this compromise proposal that involves the Clay-Bearing unit and the Sulfate-Bearing unit awaited further exploration and analysis of those units. The Siccar Point group concept places creation of the Murray formation in the late Noachian Period and creation of the Stimson formation in the Hesperian Period. The contacts between the two are irregular, indicating a significant time between the end of the deposition of the Murray formation and the beginning of the deposition of the Stimson formation, time enough for the observed significant erosion of the underlying Murray formation in some are.

The Hesperian Period is known for volcanic eruptions in the region of Hesperia Planum, with injection of massive amounts of water and sulfur dioxide into the atmosphere, resulting in rains of sulfuric acid. This may have started further reactions that resulted, after flooding, in the deposit of the Sulfate-Bearing unit on Mount Sharp. This was probably calcium sulfate, which has been seen frequently by Curiosity as veins in fractures of rocks of lower Mount Sharp. Where the calcium was picked up and how it was transported to the sulfate section of Mount Sharp is an interesting question. Possibly it was already there, on upper Mount Sharp, ready for the Hesperian rain of sulfuric acid.

Mount Sharp Chemical Factory If a core of upper Mount Sharp had been rich in calcium carbonate, then a rain with sulfur dioxide and water, might generate the calcium sulfate. Further, rain erosion might create the clay and drain down through the Murray and Stimson formations to help cement them. Calcium carbonate is common in limestone, which is often generated indirectly by lifeforms such as clams and coral. Alternately, anorthite is a calcium mineral that cam be found in crustal plagioclase feldspar from volcanos. Since anorthite contains aluminum and silicon, a raining mixture of water, sulfur oxide, and sulfuric acid might produce a river of water saturated with calcium sulfate and carrying clay. The mixture would drain down from the Sulfate-Bearing unit into the Clay-Bearing unit, precipitating clay. From there the saturated water could drain to the Murray and Stiimson formations below. The calcium sulfate could be the cementing agent that lithified the sandstone and mudstone units as they dried in arid periods.

Alternating Humidity and Aridity The Siccar group at Mount Sharp provides a remarkable opportunity to examine the transition between the contrasting environments of the late Noachian and the early Hesperian periods. It affects part of the understanding of the geology of Mars that could not have come from further orbital reconnaissance. In-situ exploration is usually necessary to determine the history of surface geology, supplemented by critical information from orbit.

Unfortunately, even though surface exploration can determine that the contact between the Murray and Stimson formations is non-conformant, which means that there is an uncertain time between their emplacements, it does not give an absolute number to that time interval. In the present state of the art for orbital and surface robots, that must await sample returns, as has been done for Earth's Moon with both robotic and human missions.

As the Mars atmosphere lost its water to space and the surface to subsurface seepage, the climate turned more arid. The surface submitted to aeolian (windborne) erosion and dune forming conditions. Erosion of the drying sediment left below Gale Lake could have supplied sand for the Stimson dunes; the dust from the amorphous component would likely have blown away. Ground water and intermittent climate change due to tilt of Mars' axis may have intermittently brought less arid conditions, turning dunes to cross-bedded rocks. Rain water on the sulfate rocks may have helped dissolve calcium sulfate and drain it down as groundwater to help cement the Stimson dunes into rock.

Water Flow Model Much of the creative thinking for the very successful Siccar group concept and its implications is based on qualitative reasoning. In a 2017 abstract for the Fourth Conference on Early Mars in October, 2017, D. G. Horvath and J. C. Andrews-Hanna presented a quantitative model for water flow for Gale crater and the surrounding region. The time range is for the Noachian Period, before 4.1 billion years ago (bya) and the Hesperian Period (from 4.1 to 2.9 bya). The model was based on the topography of Gale crater and the surrounding region, as measured from stereo photos from orbit. Assumptions were made for average rainfall, runoff, evaporation, and groundwater head (elevation of subsurface water table). The late Noachian Period was assumed to have a semi-arid climate, allowing the surface of Lake Gale to rise to the elevation of the highest Murray formation mudstones. For the results of the simulations, see Fig. 8.2.

The simulations showed different environmental conditions, characterized as semi-arid (like the Great Plains), arid (Arizona) and very arid. In the most humid case, the northern ocean surface level was high enough to overrun the northern rim of Gale crater. This condition could have occurred either with increased rainfall, such as from water released by volcano eruptions in the Hesperian Period or, at lower levels of the northern ocean, a tsunami caused by a meteor striking the ocean.

The groundwater contours show that there could be a continuing seepage of groundwater into Gale crater from the surrounding plateau to the south and from the northern ocean in Borealis Basin to the north. The level of Gale Lake, to reach the altitude of the contact of the sulfate section with the clay would still be 4,000 m below the mean Mars surface.

This complex story reported here and published in peer-reviewed journals by many authors came about because it was needed to explain the observations of Curiosity in its first 4 or 5 years of exploration at Gale crater and verified by the continuing years of exploration. There are immense assumptions and questions in this story. How is sediment brought into Gale crater by flood water over its rim carried to Mount Sharp? Where does the calcium sulfate come from? Where and for how long was it habitable? There will be more data from Curiosity's explorations and a great deal more discussions and simulations in the future. This complex explanation, spanning billions of years and the surface of a remote planet (not just Gale crater) is a good start.

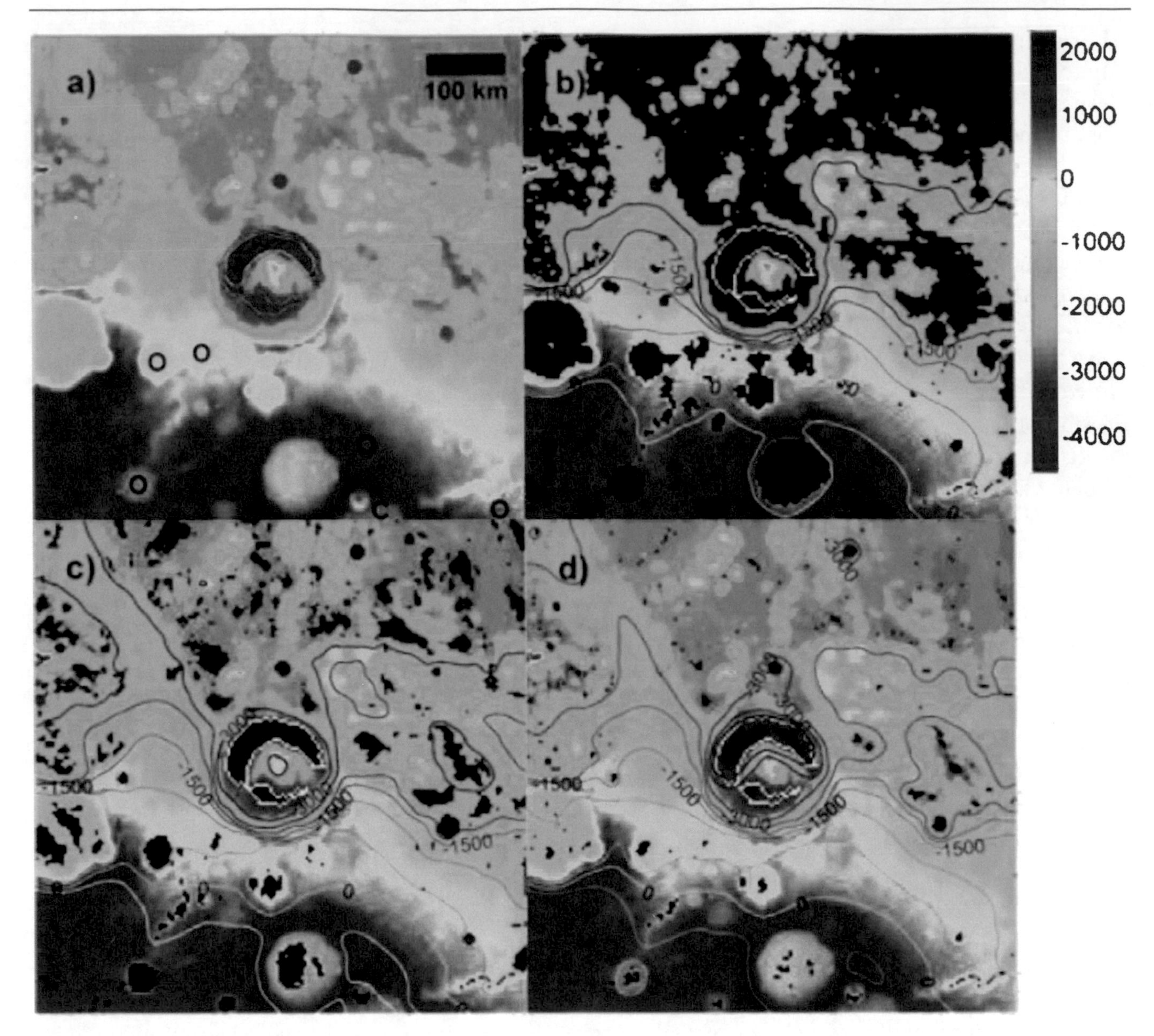

Fig. 8.2 (**a**) Topographic map of Gale crater and surrounding region. The other three maps (**b–d**) show lakes (black) and hydraulic head contours overlaid on the topography for humid, arid, and very dry conditions. This set of images is based on a digital elevation map from NASA/Mars Global Suveyor, MOLA, and the MOLA Science Team. The overlays are by D. G. Horvath and J. C. Andrews-Hanna, shown in the paper "Reconstructing the paleo-climate and hydrology of Gale crater, Mars in the late Noachian and Hesperian Epochs," AGU Geophysical Research Letters, Volume 44. Issue 16, August 15. 2017 (free access)

9 Ascent to the Vera Rubin Ridge

So, here we are, having completed the Bagnold Dune Campaign after learning a great deal about sand, cross-bedding, the Murray formation, the Stimson formation, and the Siccar Point group. Curiosity, guided by its cameras' images and the operations team planning, proceeded to the south and upward.

Vera Rubin was an astronomer who gathered data about rotation in galaxies outside our home Milky Way. Her measurements supported the existence of dark matter, whose gravity was needed to hold the galaxies intact. The ridge that took her name (unofficially so far) is in a segment of a ring on the northern side of Mount Sharp. The ridge was observed from the spectroscopic camera CRISM of the Mars Reconnaissance Orbiter to have a spectral signature of hematite, an oxide of iron. Like Calcium Sulfate, hematite can cement the grains of sand and harden sediment into rock. This may have formed a cap to protect the underlying strata from eroding as quickly as its surroundings, forming a ridge. The ridge had been named the Hematite Ridge before it was renamed to honor the contributions of Vera Rubin. Exploration of this ridge was selected as a high priority science goal, partly because it could provide a comparison between orbital observation and observation from the surface.

Like most ancient ridges, the Vera Rubin Ridge has a hazardous talus slope along its northern and southern edges. Therefore, the route to climb to the top of the 50 m ridge had to be carefully chosen (see Fig 9.1). The route would be in the Murray formation.

The bases of units of the Stimson formation have elevations below the elevation of Ogunquit Beach. Most units of the Stimson formation that may have existed along Curiosity's further route have been completely eroded away. However, there are a few mesas, buttes, and outcrops above Ogunquit Beach that may be survivors of the Stimson formation.

There was some urgency to make up time and get on up the mountain. The Bagnold Dune Campaign had taken quite a long time. Further, the drill had not worked since sol 1536, so a lot of time was spent with diagnostic tests. So a new efficient mobility plan was needed, called "Touch and Go. The early part of each sol would be for contact science where it was, and then there would be a short drive with most of the remaining power. Images would be transmitted for navigation as satellite relays became available. On the next sol, there would be more context science, then another drive (based on the new images), etc. The object was to keep going to the next high priority science objective while keeping track of the situation and being alert for targets of opportunity. This mode was used for the traverse to the Vera Rubin Ridge. Of course, atmospheric and environmental sciences were not neglected.

One factor in the decision to move expeditously toward Vera Rubin Ridge was that the ability to drill was not available because of a serious mechanical failure. Therefore, both time and power were available for mobility. During the hiatus in sampling and SAM operation, the drill engineers devised a radical work-around that required intensive testing on Earth.

C. J. Byrne, *Travels with Curiosity*, https://doi.org/10.1007/978-3-030-53805-7_9

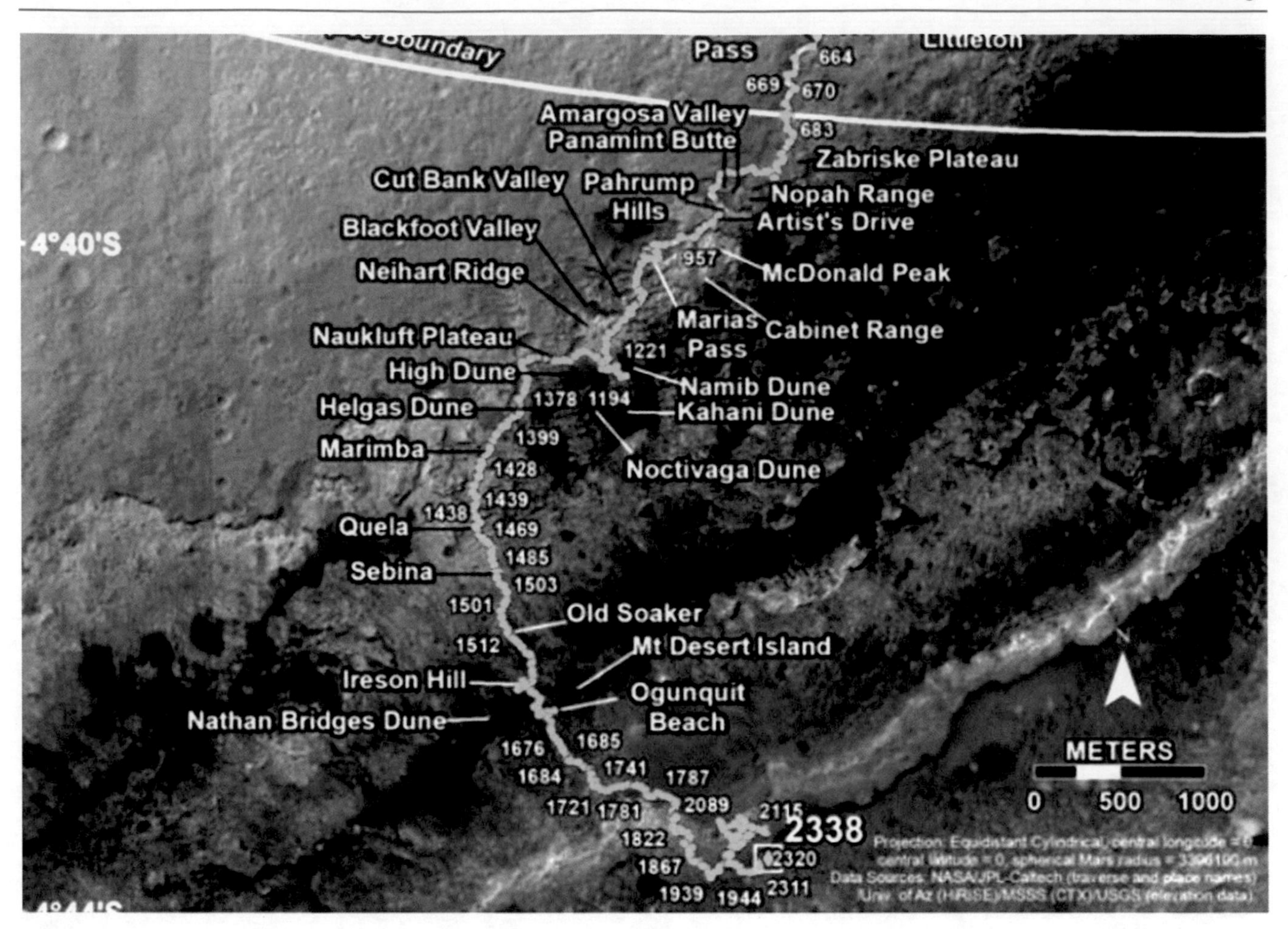

Fig. 9.1 Map of Curiosity's route from Ogunquit Beach to Vera Rubin Ridge ran south-southeast. Curiosity then turned a sharp left to go along the ridge to find a safe route to ascend (Image courtesy of NASA/Caltech-JPL, for traverse and place names, Univ. of Arizona (HIRISE) and MSSS (CTX) for base image.)

On the drive upslope from the southern arm of the Bagnold dunes Curiosity encountered eroded rocks (Fig. 9.2), possibly of the Murray formation. Normally such rocks have a flat surface with polygonal fractures of similar dimensions as these rocks. If they were eroded by water flow, the fractures may have guided the erosion, perhaps dissolving veins of minerals lining the fractures. The sand of the Bagnold dunes stretched along the base of Mount Sharp, possibly obscuring water courses similar to those explored at Yellowknife Bay.

As Curiosity continued directly to the south, on what is called the ascent route, it paused to take this image of the Sulfate-Bearing unit (see Fig. 9.3) in oblique sunlight. The shade and shadows of this section clearly show water erosion as rain fell on Mount Sharp after it formed. The runoff in this northern sector of Mount Sharp fell into the Clay-Bearing unit, which slopes down to the west, forming a river there. The Vera Rubin Ridge served as a southern bank for that river.

Curiosity has often found that calcium sulfate fills veins in the rock of the Murray formation, rising along lines of fracture. This familiar mineral, commercially called gypsum, is soluble in water; that could explain all the valleys in this picture. It appears that water saturated with calcium sulfate ran down Mount Sharp from the Sulfate-Bearing unit, saturated groundwater, and rose in fractures to make the white lines in the rocks. Calcium sulfate can also be a cementing agent, bonding grains together as sediment turns into rock.

Fig. 9.2 Shortly after leaving Ogunquit Lake (sol 1664), Curiosity took this interesting picture of Murray formation rocks eroding. Where there are usually fractures between plates, there is now erosion by either water or wind, possibly aided by sand or other sediment. Note the striations in the rocks, visible after erosion. Curiosity took this image with its Navcam after stopping 2 m into the drive because of a sensor detecting an excess angle in its suspension system, probably due to moving over similar rocks. (Image courtesy of NASA/JPL-Caltech)

Fig. 9.3 Looking ahead in just the right lighting, Curiosity took images of the sulfate unit of Mount Sharp as it continued to ascend on sol 1675. Frames for this mosaic were taken with the right Navcam. (Image courtesy of NASA-JPL-Caltech)

Curiosity moved into a section of the Murray formation that appears to have been eroded, perhaps by flowing water carrying debris (see Fig. 9.4). This is one of a series of images suggesting a strong flow of water carried down from the Sulfate-Bearing unit. It appears that a channel, obscured now by sand, once carried water ahead of the camera (West) and then down-slope to the North.

As Curiosity moved further up Mount Sharp, it was seeing more signs of water having eroded the rocks. The image in Fig. 9.5 was taken by Curiosity's left Navcam on sol 1685. It showed sand with deep ripples at a long wavelength. The deep sand was bounded on the east with a straight scarp, which may have been a bank of a steam, flowing straight downhill, at about a 15 degree angle. At that angle, flowing water would have considerable force, enough to tumble that loose rock from uphill.

Fig. 9.4 Deep megaripples of sand were bounded by a straight "bank" (of an ancient stream?) or a "scarp" (of a fault?). The megaripples of sand enhance the illusion of a flowing stream but also might simulate an actual ancient stream. Curiosity took this image of a rock called Sugarloaf on sol 1685 with its left Navcam. (Image courtesy of NASA/JPL-Caltech)

The ripples in the sand were probably due to strong downhill winds. The sand ripples essentially simulated earlier waves in flowing water. If the sand indeed occupied an ancient streambed, the stream would have flooded in a wet period and the sand accumulated in an arid period.

Moving uphill, Curiosity found this sand with ripples that look like breaking waves (Fig. 9.6). This probably implies relatively shallow sand, just as waves crest near the shore.

Curiosity came up to the ridge and turned left to follow the path up to the top of the Vera Rubin Ridge. Pettegrove Point is a projection of the ridge top, leaving some of the layers projecting out, like balconies (Fig. 9.7).

The panorama mosaic in Fig. 9.8 is of the northern scarp of the Vera Rubin Ridge (VRR). The mosaic was made from frames of Curiosity's Mastcam. Curiosity approached the VRR directly to the right side of the mosaic and then turned east to run along the ridge toward the notch in the edge of the VRR (left side of Fig. 9.8). The notch may be a landslide that allowed Curiosity to climb to the top of the ridge. Fig. 9.11 has a closer view of the notch and nearby cliff.

Climbing a landslide or a talus slope is risky. First, there are large rocks that could damage the bottom of Curiosity, with a 30-cm clearance. Second, a rock could be loose, causing a slip. Fortunately sand has infiltrated the rocks, tending to stabilize them.

Fig. 9.5 Curiosity took this image with its front Hazcam on sol 1691. The Murray formation rocks appear to have been eroded by water flow, perhaps drained down from the Sulfate-Bearing unit. (Image courtesy of NASA/JPL-Caltech)

A careful study of the rocks here was made by F. J. Calef III et al., "Rock Hazards identified from orbit approaching the Vera Rubin Ridge, Gale Crater," LPSC 2018. Using HiRISE images from the Lunar Reconnaissance Orbiter in combination with in-situ measurements, the sizes of individual boulders were estimated from their shadows on the HiRISE images. Measurements were made of the number of rocks deemed to be hazardous in 10 m squares. Sections of the original path, that had been routed beyond the talus slope at the northern foot of the Vera Rubin Ridge was found to have a high rate of hazards (see Fig. 9.9).

Apparently most large rocks and boulders eroding from the scarp of the ridge had rolled or bounced off the talus slope and fallen beyond its toe. The hazardous section of the path was re-routed towards the toe of the talus slope. Curiosity successfully traversed the revised path to the top of the ridge.

Fig. 9.6 Here are more long wavelength ripples, but these appear more like breaking waves. Curiosity dipped her wheels in the sand to check the traction and depth. The image was taken on sol 1751 by Curiosity's right Navcam. (Image courtesy of NASA/JPL-Caltech)

Curiosity came up to the ridge and turned left to follow the path up to the top of the Vera Rubin Ridge. Petttegrove Point (Fig. 9.7) is a projection of the ridge top, leaving some of the layers extending out, like balconies.

Curiosity took a long-range view of the ridge as it moved along it toward the notch (see Figs. 9.10 and 9.11).

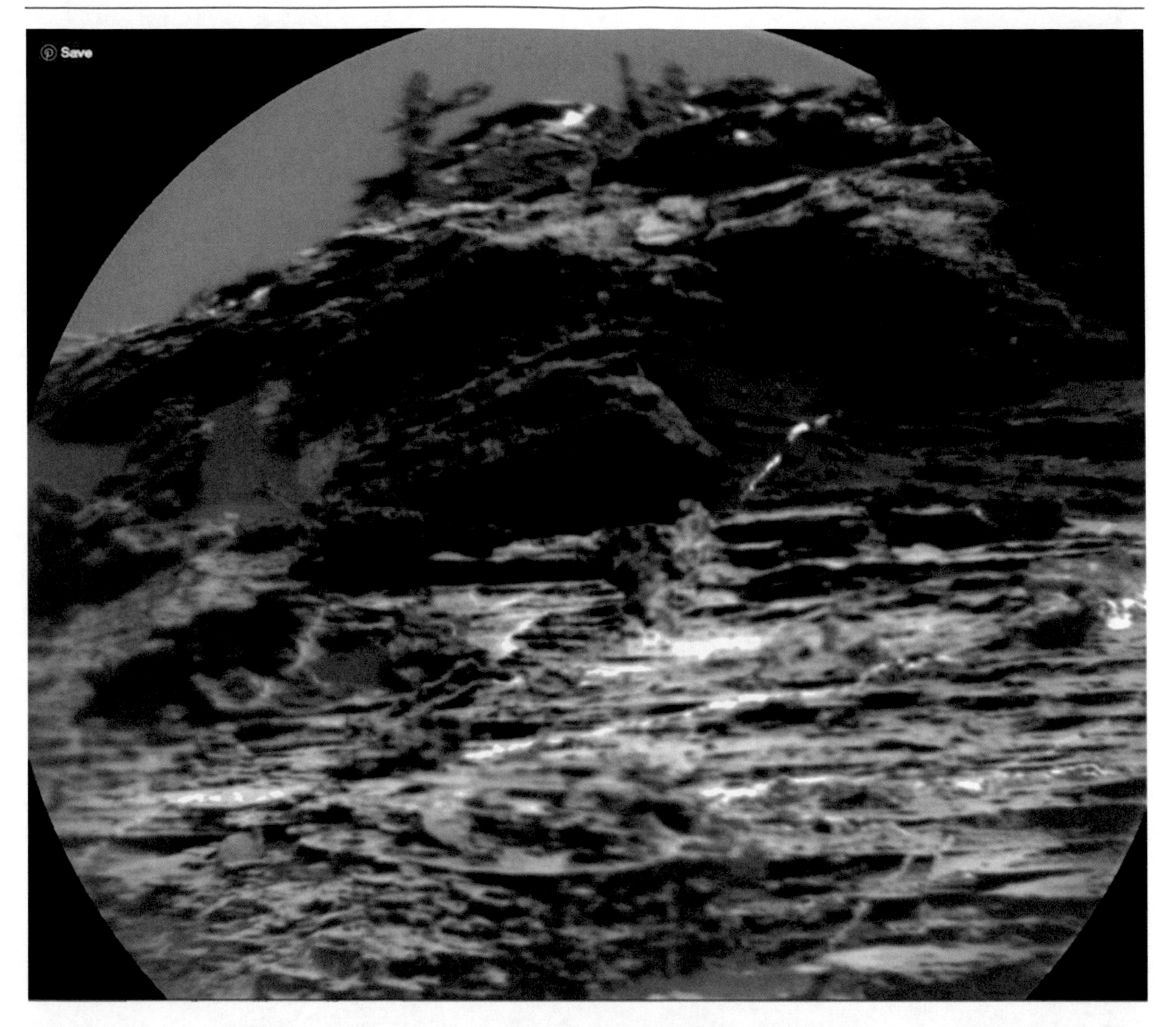

Fig. 9.7 Pettegrove Point, a promontory on the edge of Vera Rubin Ridge, as seen from below. Curiosity took this picture with its ChemCam on sol 1795. It shows the overhanging rocks and the striated base of Vera Rubin Ridge. Note the white streaks in the veins, probably calcium sulfate. (Image courtesy of NASA/JPL-Caltech/LANL)

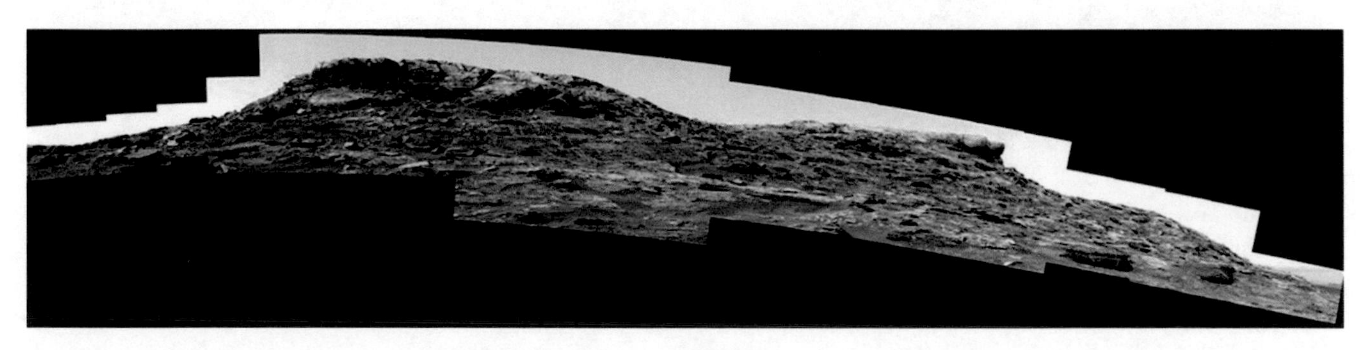

Fig. 9.8 This view of the northern scarp of the Vera Rubin Ridge is west of the notch that Curiosity was to climb to get to the top of the ridge. Note the large slab of rock that has fallen from the cliff left of center and lays on the top of the talus slope, leaving a light-colored scar behind. On a map, the scarp is curved; the center of the two prominandes bends away from the camera. This image shows erosion in progress. (Image courtesy of NASA/JPL/Caltech/MSSS)

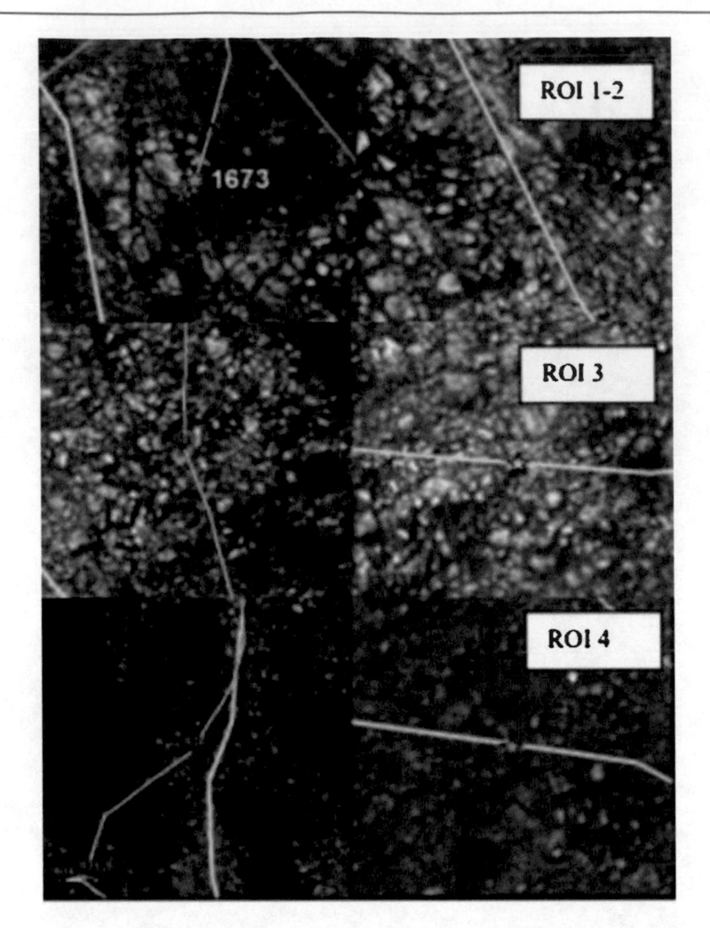

Fig. 9.9 Sections of HiRISE images of alternate paths for Curiosity to approach the Vera Ruben Ridge. Qualitative comparison of rock density: previously visited terrain on the left and regions of interest (ROI) on the right, at the same scale. The revised path was more like ROI 4k, replacing a section like ROI 3. (Image courtesy of Calef et al., LPSC 2018 Abstract 2510, Figure 6)

Fig. 9.10 This is a mosaic of Freeman Ridge from Curiosity's Mastcam on sol 1726. She was at waystop and rock Jones Marsh, the beginning of the Vera Rubin Ridge campaign. (Image courtesy of NASA/JPL-Caltech/MSSS/Thomas Appere)

Figure 9.12 shows Curiosity beginning to start up the steep path to the VRR. The path runs over what appears to be a landslide from a collapse of the edge of the VRR. It slants upward to the southeast, so Curiosity looks up toward Pettegrove Point as it goes up the slope.

Fig. 9.11 Curiosity took this detailed picture of the Vera Rubin Ridge to the west of the notch she would use to ascend to the top. This is a close-up of the rock structure in Fig. 9.8. Curiosity took the image with her Mastcam on sol 1800. (Image courtesy of NASA/JPL-Caltech)

Perhaps this long Labor Day weekend, you will find yourself on a hike that ends with a particularly steep stretch. On such a steep trail, you will simply want to concentrate on reaching the top of your destination and not be required to perform any other tasks. Imagine, instead, you were asked to dribble a soccer ball, juggle and sing a show tune while making your last push up the hill. Now you know how Curiosity feels this weekend! In addition to simply driving up the 20 degree slopes on the flank of Vera Rubin Ridge, Curiosity will acquire an amazing variety of science observations of the ridge rocks.

The update blog for the image of the start was posted by Michelle Miretti on September 5, 2017. It describes just how complex Curiosity is:

Robots do not tire, like people, but their operators must consider the level of power available in their short-term power source, usually a battery. Curiosity will shut down if it reaches too low a charge, but an operator does not want that to happen if Curiosity is in a precarious position.

The start of the path up the slope of the notch (Fig. 9.13) appears to be covered by the plates of the Murray Formation, but then the plates have been heavily eroded. Some laminations have been broken off

Fig. 9.12 Up to Vera Rubin Ridge! This is the start of the path up to Vera Rubin Ridge. The opportunity to rove up here appears to be due to a landslide event of the northern scarp (cliff) of VRR. The "paving stones" are rocks of the Murray formation. The top layer of sediment appears to be stripped off, perhaps by the landslide debris. This image was taken by Curiosity's left Navcam on sol 1802 and shows its logo on the stowed arm. (Image courtesy of NASA/JPL-Caltech)

Fig. 9.13 Curiosity took a picture of the edge of the northern scarp of the ridge while climbing the slope towards the top of the notch (see also Figs. 9.8, 9.10 and 9.11). Curiosity took the two images of this mosaic with its right Navcam on sol 1803. Mount Sharp is beckoning, across the flat slope of the top of the Vera Rubin Ridge. (Image courtesy of NASA/JPL-Caltech)

the tops of the plates and there is quite a bit of rubble strewn about. The impression is that the wall formed by the top of the northern scarp being higher than the flat surface of the VRR has collapsed, forming the notch and the slope. Then the Murray Formation sediment may have been laid over the slope, partly smoothing it. Further erosion by water and windblown sand may have further smoothed the slope.

As Curiosity went up the slope, she took a picture of her next goal, the top of the Vera Rubin Ridge. The broken edge of the northern scarp of the VRR rises higher than the edge of the southern flat at the top of the VRR. That edge bounds the west side of the notch. (see Fig. 9.13).

Fig. 9.14 Curiosity is on the steep part of the slope, where there are some large rocks that could be hazardous. The view is to the left of Curiosity, showing the top of the northern scarp of the Vera Rubin Ridge; these rocks are on the eastern side of the notch. She took the picture on Sol 1807 with her right Navcam. The slope that Curiosity is climbing appears to be a landslide from the edge of the top of the Vera Rubin Ridge, where the rocks have tumbled. Wind, water, and sand between rocks have leveled this slope somewhat. Aeolis Palus stretches toward the northeastern edge of the Gale crater, about 300 m below Curiosity's position. The eastern side of Mount Sharp is on the horizon. (Image courtesy of NASA/JPL-Caltech)

Fig. 9.15 The width of the ridge is divided into four sections, the lower ridge, middle ridge and upper ridge and the northern ridge. This can be described as the edge of the lower ridge. Now we can see the face of the rock, which looks very much like the Murray formation (Image courtesy of NASA/JPL-Caltech)

Nearing the top, Curiosity took the image of Fig. 9.14, taken on the central part of the trail. The image shows how sand has partly leveled the rocks. There is a lovely view of eastern Aeolis Palus beyond the edge of the slope.

To take the next photo, Curiosity had to take a position at the very edge of the notch (see Fig. 9.15). She is now at the top of the slope, at the level of the top of the Vera Rubin Ridge, looking along the top edge of the northern scarp. It is interesting that the Murray formation appears to be rolled over the edge. This could be true if erosion had eaten away the ridge underneath the upper slabs of rock and then they folded down like a piece of cardboard or it could also be that the upper layers were eaten away in a rounded fashion and then one or more new layers of Murray formation were deposited on the eroded face.

Curiosity had shown us a very similar image at a much lower elevation, at the Big Sky location. There, the edge could have been the downwind edge of a Stimson formation barchan dune that was being eroded, covered by a new Murray formation. In any case, how the Murray formation can drape itself over a rounded edge is a puzzle.

Fig. 9.16 With its wheels firmly on the Vera Rubin Ridge (lower right corner), Curiosity looked back all the way from Vera Rubin Ridge to Bradbury Landing, where it started on Mars. The north rim of Gale Crater is in the background. Curiosity's Mastcam took the frames for the mosaic. (Image courtesy of NASA/JPL-Caltech/MSSS)

Looking at the relatively level part of the VRR in (Fig. 9.15), it looks very much like the Murray formation. The science team needed to see drill samples analyzed by SAM to find out what was different about the ridge. Unfortunately, the drill had not been available for over a year because of a broken part, the feed mechanism. A lab-tested workaround was now available, so the field tests were conducted on Vera Rubin Ridge.

Curiosity, securely on Vera Rubin Ridge, looked back all the way to its start at Bradbury Landing. The remarkable panorama in (Fig. 9.16) shows Curiosity's landing point against the background of Gale crater's northern rim. Most of the territory covered in its first year in Gale Crater was visible, including Yellowknife Bay and the traverse toward Mount Sharp beyond Cooperstown (Fig. 9.16).

In general, ridges are either tectonic in origin or, like buttes, have top layers that have relatively resistant to erosion and therefore protect the underlying layers. The working hypothesis was that the rocks on the top of the ridge had more hematite in their composition than neighboring rocks, and that was the reason the ridge remained while other areas eroded. The first objective was to confirm that the quantitative orbital data correlated with in-situ (on the surface) data. The second objective was to determine if the surface layers are indeed unusually hard and establish why that is so.

Since the orbital images indicated that the surface attributes changed more with width than to length along the ridge, the first goal of the walk-about was to cross the ridge. This had the secondary value of bringing the clay area on the far side into Curiosity's view early in the walk-about, which gave the navigation team more time to plan how to get down to it. The second part of the walk-about was to turn at a right angle to get a feel for variation in that direction.

The top surface of the Vera Rubin Ridge is covered with Murray formation rocks, somewhat eroded and often covered in sand and debris. So in order to sample, it was necessary to seek out bed rock exposures with rocks stable enough for drilling and that could be cleaned with the brush, if possible.

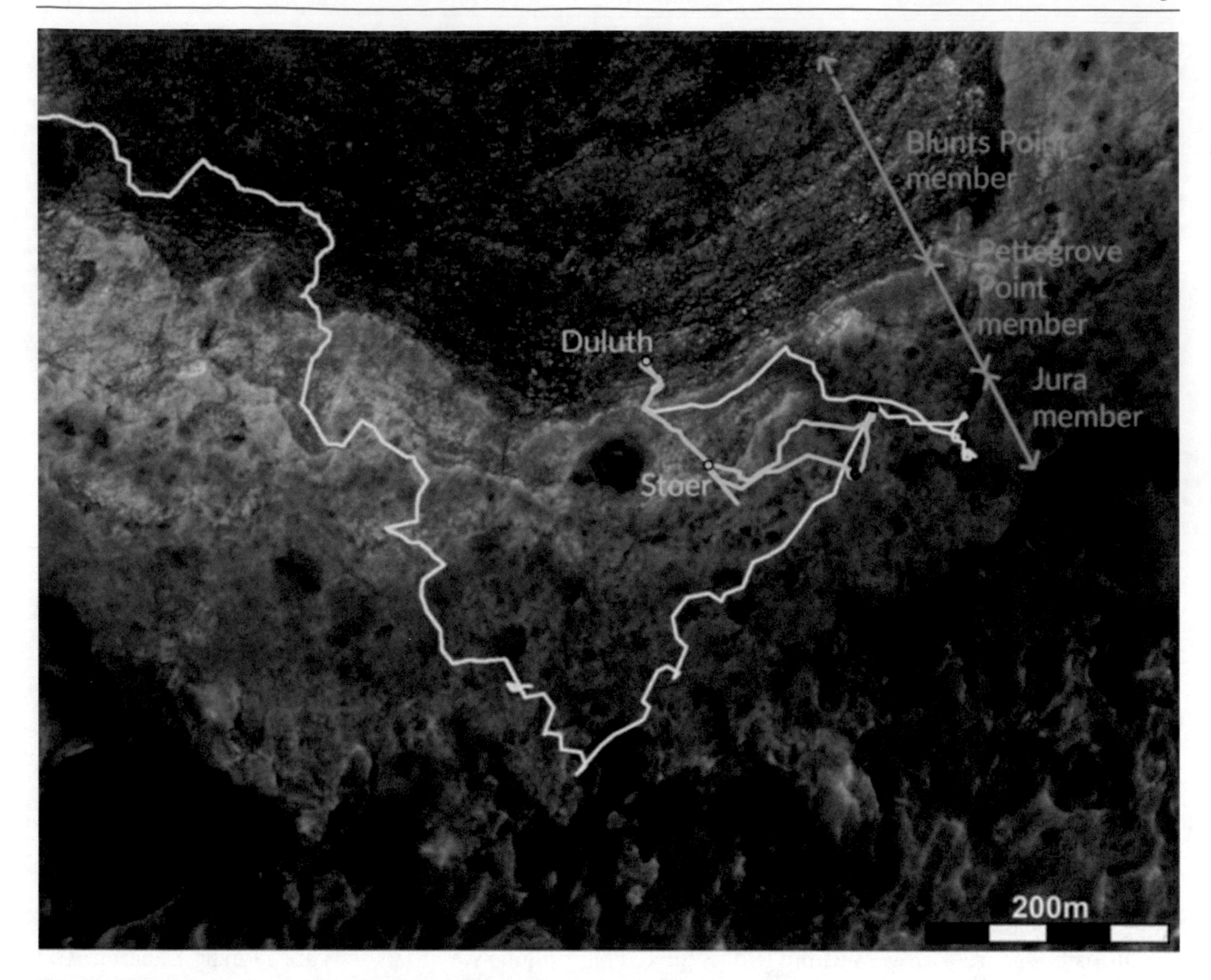

Fig. 9.17 This map shows the walk-about route (white) and the subsequent partial contact science route (yellow) for the exploration of the Vera Rubin Ridge. Three new units of the Murray formation were identified: Blunts Point member, Pettegrove Point member, and Jura member. This base photograph is from HiRiSE, the high-resolution camera of the Mars Reconnaissance Orbiter. (Image courtesy of NASA/JPL-Caltech/Stooke/Lakdawalla)

After the walk-about, Curiosity was directed to areas of sufficient interest to do contact science, including, if possible, drill samples sent to SAM. The goal was to obtain four samples from the ridge, each representative of areas with different compositions, if possible. If it would not be possible to recover the ability to drill, remote instruments might meet the objectives.

The maps in Figs. 9.17 and 9.18 show the initial walk-about, whose objective was for the science team to get a feeling for the general situation at the surface units of the Vera Rubin Ridge. The first map has the highest evolution and the second map shows the relative amount of the hematite mineral by spectroscopic analysis from orbit. The maps are derived from an interactive map on Emily Lakdawalla's blog that used Phil Stooke's Curiosity route.

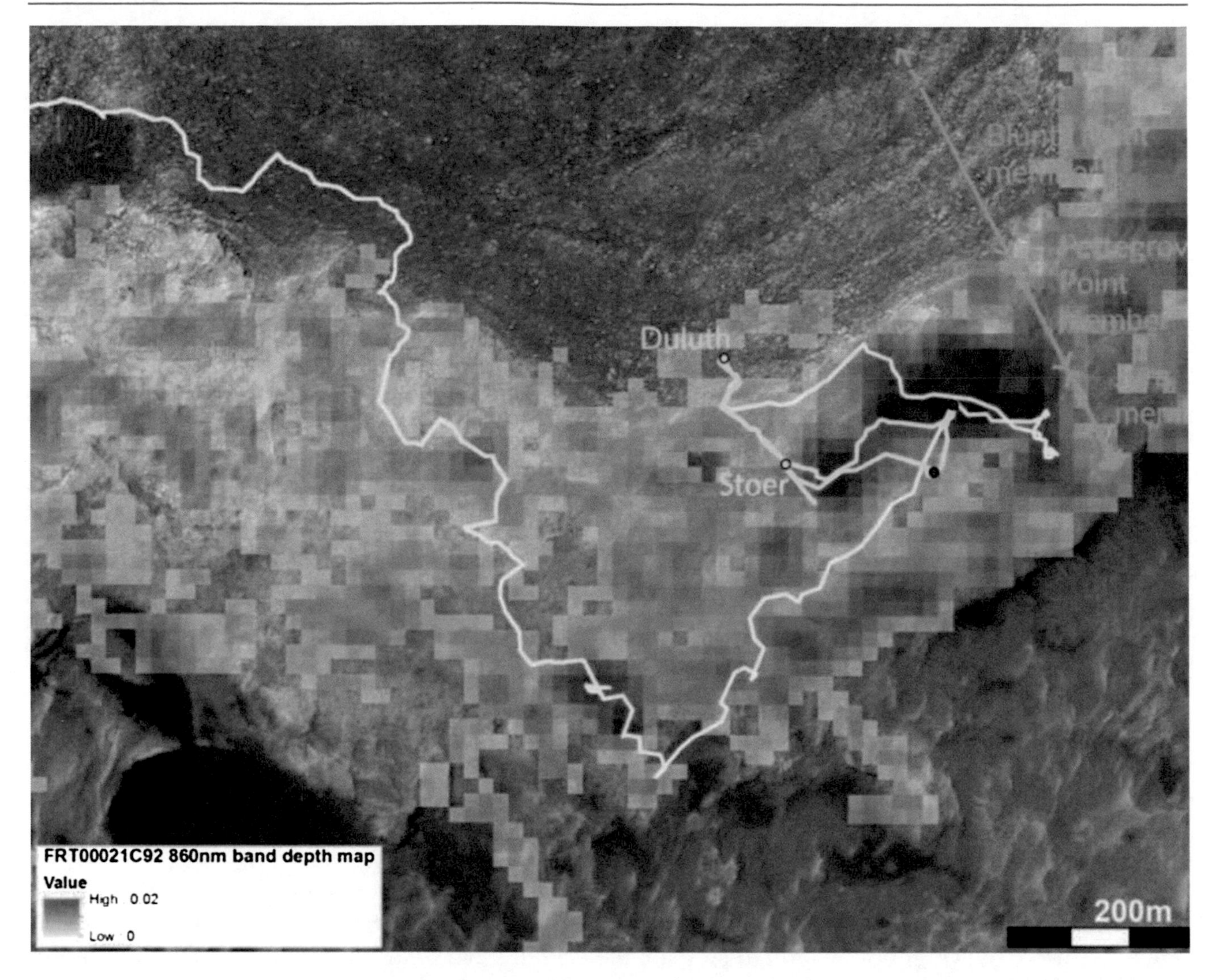

Fig. 9.18 This map shows the walk-about route (white) and the subsequent partial contact science route (yellow) for the exploration of the Vera Rubin Ridge. The base map is from the CRISM spectrometer of the Mars Reconnaissance Orbiter. It records the peak of the band depth of hematite (red). (Image courtesy of NASA/JPL-Caltech/JHUAPL/Stooke/Lakdawalla)

The white path in Fig. 9.18 shows Curiosity's approach and walk-about. The yellow represents attempts in the early sampling campaign. This image was taken from the Planetary Society web page.

In the course of the walk-about, Curiosity crossed the Vera Rubin Ridge to the top of its southern cliff. From the edge, Curiosity could look down to the bottom of the clay unit (see Fig. 9.19). It also could view the path it would later use to safely descend to that unit as well as look back at the ridge's cliff. Of course this opportunity was planned in setting the path of the walk-about.

After the walk-about, there was sufficient information to revise the sections of the Vera Rubin Ridge from lower unit, middle unit, and upper unit to Blunts Point member, Pettegrove member, and Jura member, each of the Murray formation. The Jura member had two facies, the gray Jura and the red Jura. Accordingly, the goal was to obtain one sample from each of the newly identified units, a total of four samples. This goal was accomplished, but it took well over a year to do it, including time to develop the techniques needed to use the new drilling mode and to deal with various glitches along the way, collectively called "the curse of Vera Rubin Ridge."

Fig. 9.19 Curiosity took the images for this mosaic from the southern edge of the Vera Rubin Ridge. Curiosity is near the path it will later follow to descend to the clay region. The light-colored Sulfate-Bearing unit and the peak of Mount Sharp are in the background. (Image courtesy of NASA/JPL-Caltech/MSSS)

The drill had not been used for over an Earth-year because the drill extending device broke. A work-around procedure had been designed and tested on Earth but had not been used on Mars yet. The new procedure used the weight of the arm to press on the drill, instead of the extender. Percussion, rotation, or both could still be used in drilling. Initially, the engineers were reluctant to use percussion to avoid further breakage but it turned out both were needed because of the hardness of the Murray foundation bedrock on the Vera Rubin Ridge.

Two attempts were made at sites selected after the walk-about. At the first site, Stoer in the Pettegrove Point member tried to drill with only rotation (no percussion) but succeeded in making only a shallow hole, leaving some tailings to analyze, not enough for a sample. At a second nearby site, with percussion but no rotation, the drill even did not penetrate the surface.

The science team picked a new location, over the edge of the ridge, a medium sized layered rock, called Duluth (see Fig. 9.20). This time, both percussion and rotation were used. The drill was successful this time (see Fig. 9.21), including transferring the sample to SAM. So now, Curiosity was able to attempt drilling and sampling. The process usually but not always worked. Still, this was a major accomplishment since it restored the ability to analyze samples with the SAM suite of instruments.

Having demonstrated the new drilling technique, Curiosity returned to the Stoer location in the Pettegrove member to drill a new rock close to where the drill had not been successful. This time both rotation and percussion were successful in drilling a full depth hole and collecting a full sample (Fig. 9.22).

Fig. 9.20 The science team picked out this medium-sized rock Duluth, imbedded in sand, to test the new drill process, since the rocks at top of the ridge had, as suspected, been too hard to drill. So how hard can this rock be, which looks like Greek pastry? (Image courtesy of NASA/JPL-Caltech)

Fig. 9.21 Duluth drill hole. After well over an Earth year, a full drill hole! Further, it gives access to ChemCam to the layers inside. Curiosity took the image on sol 2057. Note the thickness of the top layer of fractured rock, showing brittle failure. (Image courtesy of NASA/JPL-Caltech/LANL)

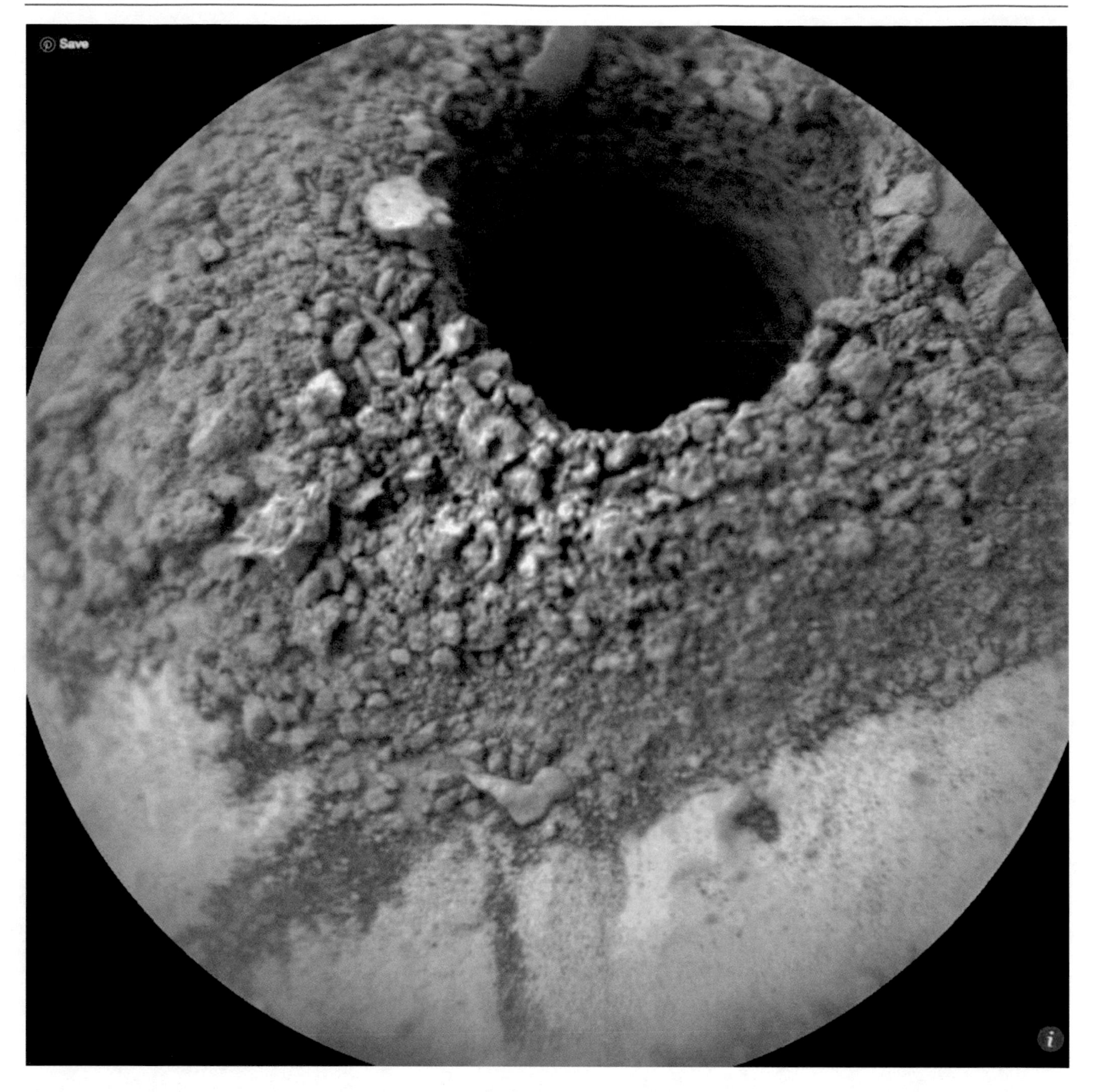

Fig. 9.22 Stoer drill hole. This shows the successful drill hole and sample collection at Stoer in the Pettegrove member of the Murray formation. The position of Curiosity relative to the rock and the lighting permit a view of the layer edges below the surface. The top layer, which generated most of the tailings, was more brittle than the deeper layers, an indication of the reason it is hard. A phenomenon called diagenesis was probably involved. After the sediment was deposited from water and dried, a second wetting of water saturated with a cementing mineral may have deposited the mineral in the pores of the sediment, hardening it. Curiosity took this image on sol 2136 with its ChemCam. (Image courtesy of NASA/JPL-Caltech/LANL)

Fig. 9.23 Highfield rock at Lake Orcadie. Curiosity had visited this area (and took this image) on sol 1962 with its front Hazcam. After two tries, its drill still could not get a sufficient sample. It returned and got a needed gray Jura sample on sol 2224, with improved drill technique. (Image courtesy of NASA/JPL-Caltech)

For the gray Jura member, the science team selected the Lake Orcadie location (see Fig. 9.23). Unfortunately, the first tries there failed, so Curiosity returned to the site after the Stoer success, to find a softer rock. This time it worked! So it was learned that some rocks are harder than others, even those in proximity to each other (Fig. 9.24).

Fig. 9.24 Gray Jura member of the Murray formation at last! This image shows the Highfield drill hole and tailings in context near the center of the image. Curiosity took this image with its left Navcam on sol 2224. (Image courtesy of NASA/JPL-Caltech)

Fig. 9.25 The bedrock slab on the right, Rock Hall, was the closest Curiosity could find to a red Jura rock to drill. The drill and sample collection were successful. Curiosity's Mastcam took this image on sol 2256. (Image courtesy of NASA/JPL-Caltech/MSSS)

The final sample the science team requested was one from a red Jura rock. They were hard to find, although the CRISM data said they should be there, a rock with lots of hematite (see Fig. 9.25).

The drill hole for Rock Hall, the representative of the red Jura member of the Murray Hill formation, is shown in Fig. 9.26. In this case, the picture was taken after the sample had been transferred to SAM. As was usually done, extra parts of the sample were dropped on the surface for examination by APXS for element identification MAHLI for grain size, etc.

Fig. 9.26 Here is the final drill hole in the Vera Rubin Ridge, its picture taken by Curiosity on sol 2263 with her left Navcam. Rock Hall was selected as an example of red Jura. (Image courtesy of NASA/JPL-Caltech/MSSS)

Results

Here are the results from Curiosity's trip to the Vera Rubin Ridge:

Murray formation (on the way up to the top of the Vera Rubin Ridge from Ogunquit Beach). Curiosity's explorations contributed strongly to understanding Martian geology, Goal II of the Mars Exploration plan. Curiosity continued to encounter Murray formation bedrocks with sample compositions very similar to each other (see Fig. 9.27). The amorphous component (mud) was always the dominant one (for sedimentary rock that's what distinguishes mudstone from sandstone). The hematite component in particular was consistent. All the way up to the top of the Vera Rubin Ridge, any exposed bedrock qualified as Murray formation, mudstone deposited in quiet water. The rocks of the Stimson formation had been left behind below Ogunquit Beach, at least on Curiosity's route.

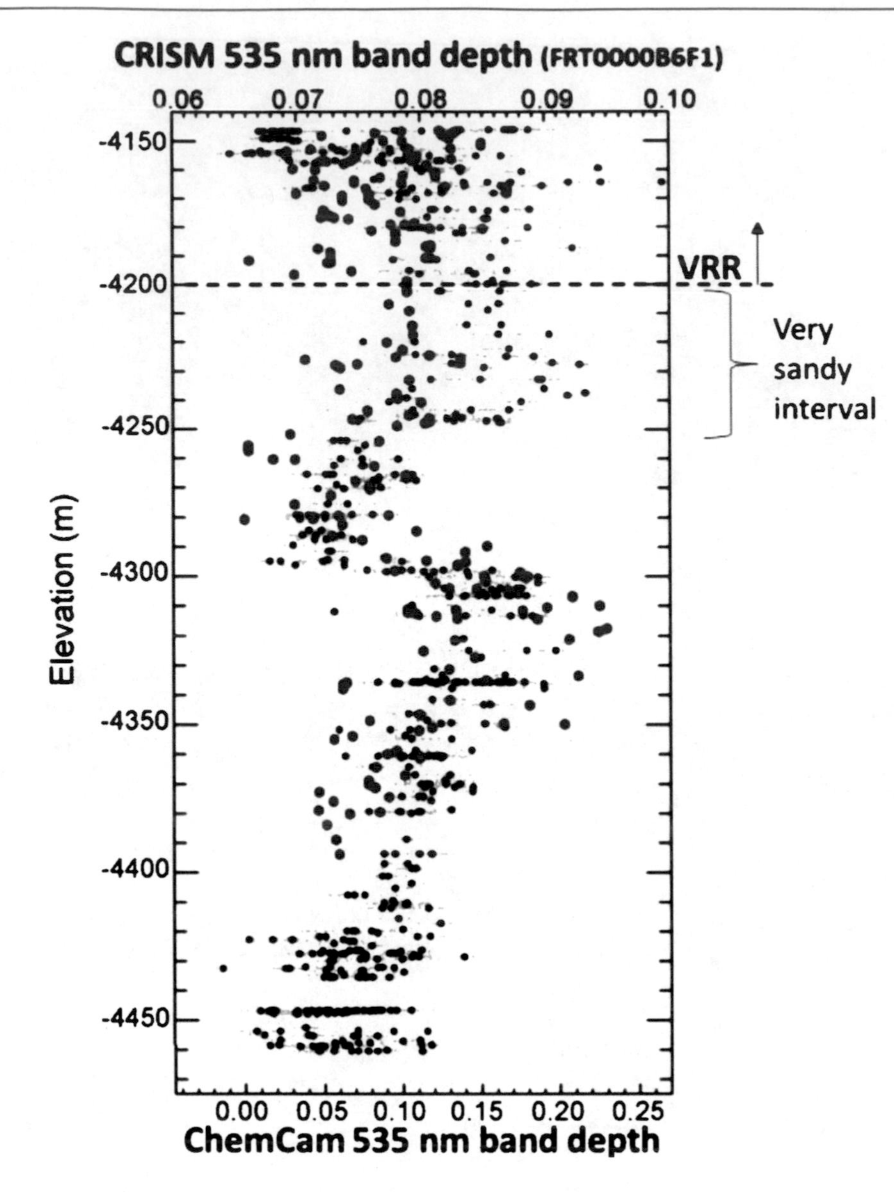

Fig. 9.27 CRISM data (red) is compared to ChemCam data (passive mode, gray and black). Hematite 535 nm band depth is plotted against elevation (relative to the mean surface). ChemCam light gray dots are for each datum, black dots are for the average for each target. The height of the ridge is about 65 m. (Image courtesy of A. A. Fraeman et al., LPSC 2019, #2118)

Curiosity had seen rocks of the Murray formation eroded by water at Yellowknife and probably Kimberly. In this chapter there are images of similar channels in the Murray formation rocks that are now filled to the brim with active sand. Two such channels run near the base of the Vera Rubin Ridge, possibly once catching water draining down from the top of the ridge along Curiosity's ascent path. After running west along the ridge, the water could have run downhill in the straight channel seen as Curiosity started up from Ogunquit Beach (Fig. 9.5). Evidence of erosion of Murray formation rocks is not unusual but it is not generally caused by wind.

Composition of Samples Drilling was attempted at several sites on or near the top of Vera Rubin Ridge and finally succeeded in collecting four samples. The first two, at the area called Stoer in the Pettegrove member, were tentative tests of the new drilling technique, first with rotation only and then with percussion only. Both attempts failed to drill deep enough to gather acceptable samples. A softer rock, Duluth, was chosen from rocks below the top, apparently eroded from the side of the ridge. Both rotation and percussion were used. This attempt succeeded (see Fig. 9.26).

After Duluth (Blunt's Point member) was successfully drilled, Curiosity returned to the top of the Vera Rubin Ridge and drilled holes for three more samples, using the combination of percussion and rotation. The samples were Stoer (Pettegrove Point member), and grayish Highfield and reddish Rock Hall Jura members. Although hematite was present, it was not greatly more abundant than in neighboring regions. What was unusual was a higher level of mudstone (amorphous material). This decrease the crystalline component, and tends to exaggerate the spectrograph peaks of hematite reported by CRISM. This was an important distinction between orbital and in situ observation. More mud in the mudstone can lead to an overestimate of the percent volume of a particular crystalline component, because the greater component of mud cannot be measured from spectroscopic observation.

ChemCam Data Curiosity measured the hematite content of Murray formation rocks at many locations of lower Mount Sharp (see Fig. 9.27). The data show that the abundance of hematite at the top of the Vera Rubin Ridge is not much greater than it is near the ridge's base. A study comparing the CRISM data from the Mars Reconnaissance Orbiter with Curiosity's ChemCam data indicated that the higher abundance of hematite reported by CRISM was partly due to its relatively large pixel diameter compared to ChemCam's laser pulse and spectrometer. Another factor was the uneven coverage of the Murray bedrock slabs with sand. Sand may have more of the crystalline material than its parent body and therefore a higher proportion of hematite. Other studies have found that amorphous dust is systematically blown away from active sand.

Rock Strength and CheMin Data Measurement of rock strength was not explicitly measured by Curiosity originally, but a kind of measurement was available in the drilling process. Percussion was applied in stages, depending on perceived need. In the new process, the percussion shocks were counted until the desired depth was reached, a more precise measure. Figures 9.28 and 9.29 show the rock strength and the abundance of calcium sulfate and hematite, respectively.

A well-known process for turning sedimentary deposits into rock is cementing. The cementing agent can be introduced by fluids wetting the material after it has dried, either by flow or groundwater. The fluid can bring the agent in solution, and subsequent drying and rewetting cycles can build up the agent. The process is called diagenesis

On the floor of Gale Crater, calcium sulfate is not common. On lower Mount Sharp, it often appeared in veins of a rock. It is likely to be the dominant mineral in the sulfate-bearing unit.

Calcium sulfate could be a hardening agent for the top layers of Vera Rubin Ridge. However, it also seems to be common, at that elevation, for its neighboring rocks. This might not produce such a major sharp-sided ridge.

In summary on this point, the top surface of the ridge was found to be harder to drill than that at lower elevations but its hematite content is about the same. Explanations for spectroscopic observations to the contrary were proposed. Sample analysis found a relatively greater percentage of amorphous material (mud) and less crystalline material in the bedrock at the top of the ridge. Further, the resolution of the orbital instrument was at best marginal for distinguishing between bedrock and sand cover on the top of the ridge. The hardness may be due to a thicker crust on the top of the rock slabs, perhaps because of longer exposure to water saturated with cementing minerals.

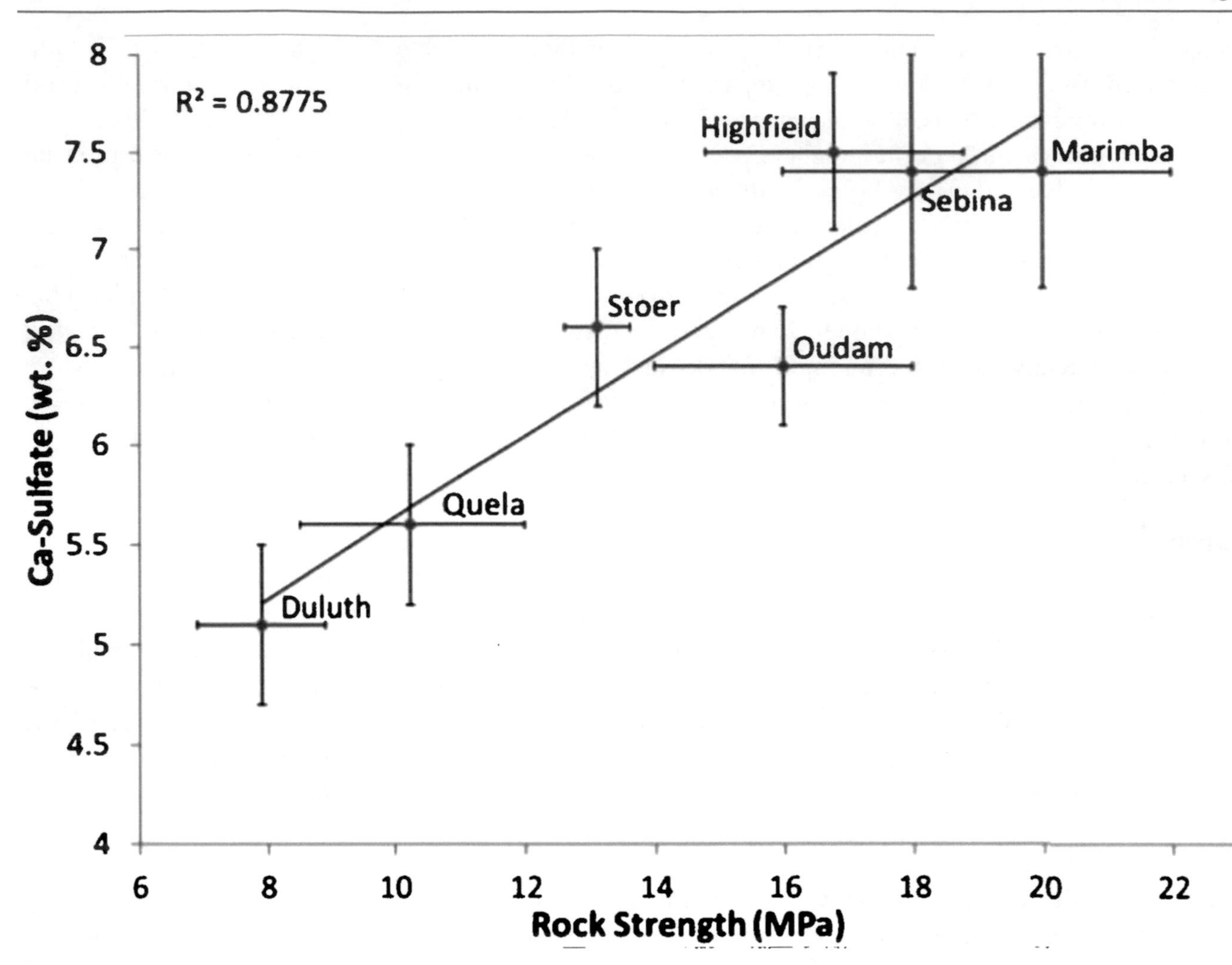

Fig. 9.28 The abundance of calcium sulfate in a sample, measured by CheMin, plotted against rock strength, derived from the rate of percussion shocks needed to drill at a desired rate. Duluth is on the shoulder of the Vera Rubin Ridge and Stoer and Highfield are on the top. The correlation indicates that calcium sulfate is a likely cement added by fluids after the initial deposit of the Murray formation for each of these sampled rocks. (Image courtesy of S. R. Jacob et al., LPSC 2019 #1671, Figure 4)

So How Was the Vera Rubin Ridge Formed? New ideas are needed to explain this ridge, now that the idea of hardening by hematite has pretty much been deprecated. Geologists are working on diageneses processes that could modify the Murray formation by interaction with additional liquids after or during the cementing process. However, an even bigger idea may ultimately explain the origin of this ridge — the Jura concept.

A hint of a radical new formation theory is suggested by the name of the Jura member of the Murray formation, in the upper level of the Vera Rubin Ridge. "Jura" is a geology term to describe similarity to the Jura Mountains near the border of France and Switzerland. There are three long, narrow ranges, each similar to the Vera Rubin Ridge. Each range is a fold in the crust caused by compression stress due to the northward motion of the Swiss Alps. The formation of each range was separated in time from the other two events. In each case, a strip of material between two joints was raised relative to the surrounding territory. A joint is a preexisting fracture, separating the raised unit from the adjoining unit.

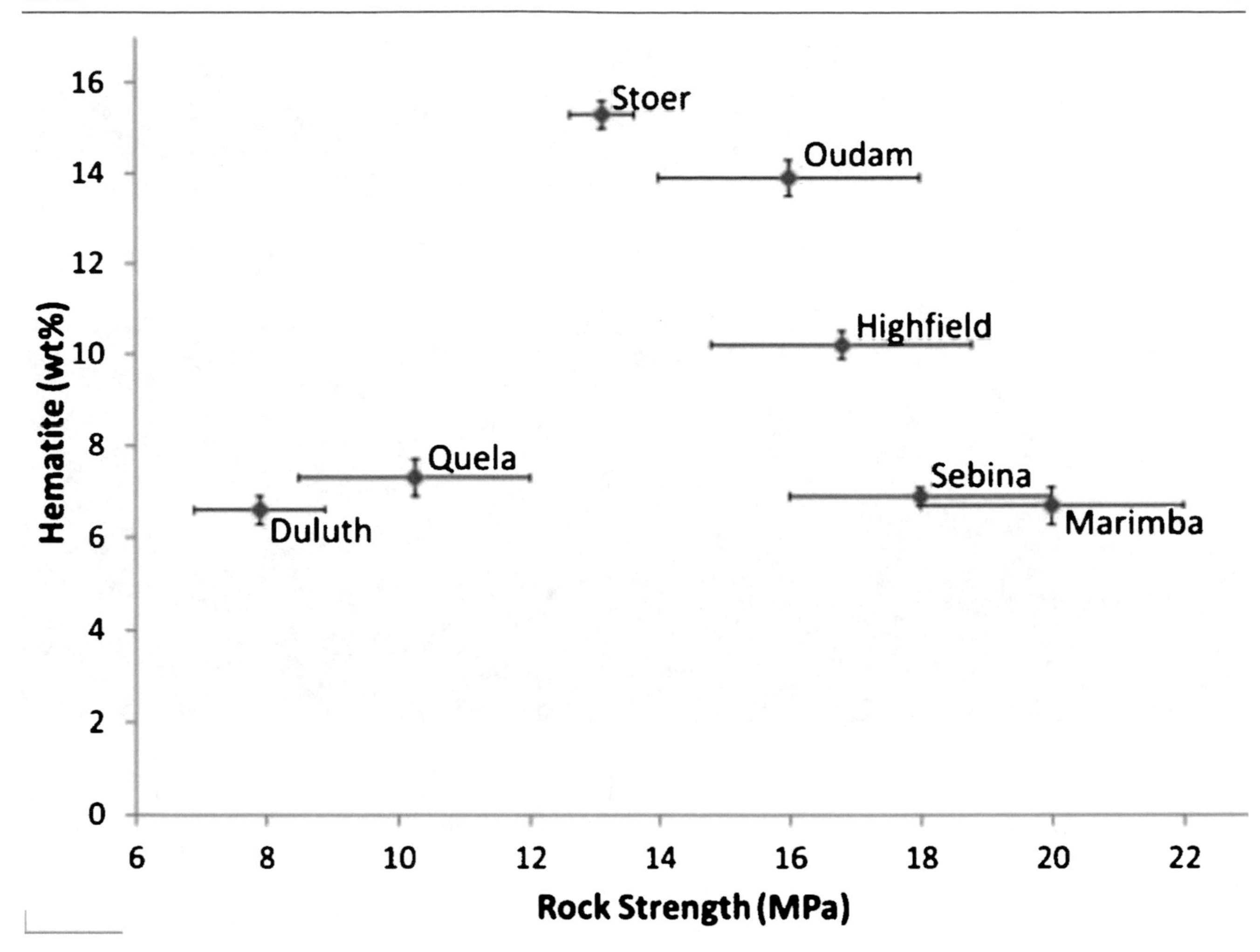

Fig. 9.29 The abundance of hematite in a sample, measured by CheMin, plotted against rock strength. The lack of correlation does not support the hypothesis that hematite abundance is a major cause of high rock strength in this case. (Image courtesy of S. R. Jacob et al., LPSC 2019 #1671, Figure 5)

Possibly, the Martian Vera Rubin Ridge could also be a raised section, perhaps with some outward movement, to relieve compression stress. The compression could be caused by an increase of mass above the ridge. Such a mass could be due to a volcanic pipe adding mass above the sulfate region. This hypothesis would permit but not require the Pettegrove Point and Jura members to be unusually hard.

The hypothesis solves other problems. For example, central peaks are common in meteorite craters the size of Gale, but their peaks do not usually rise above their rims, as Mount Sharp's peak does. Finally, if the lava receded, leaving a hollow, it might explain a recent gravity study of Curiosity accelerometer data concerning low density in Mount Sharp. (Lewis et al., Science, February 1, 2019).

Hydrology As Curiosity ascended Mount Sharp, evidence was accumulating of the great force of flooding water as she was able to get better views of the Vera Rubin Ridge in the context of the Clay-Bearing and Sulfate-Bearing units. Hydrology is the branch of geology that deals with the dynamic role flowing water plays in geology. During the Hesperian Period following the Noachian Period, the massive volcanoes of Mars repeatedly erupted. From them enormous masses of water and sulfur dioxide were released into the Martian atmosphere. This mixture created floods of acid rain.

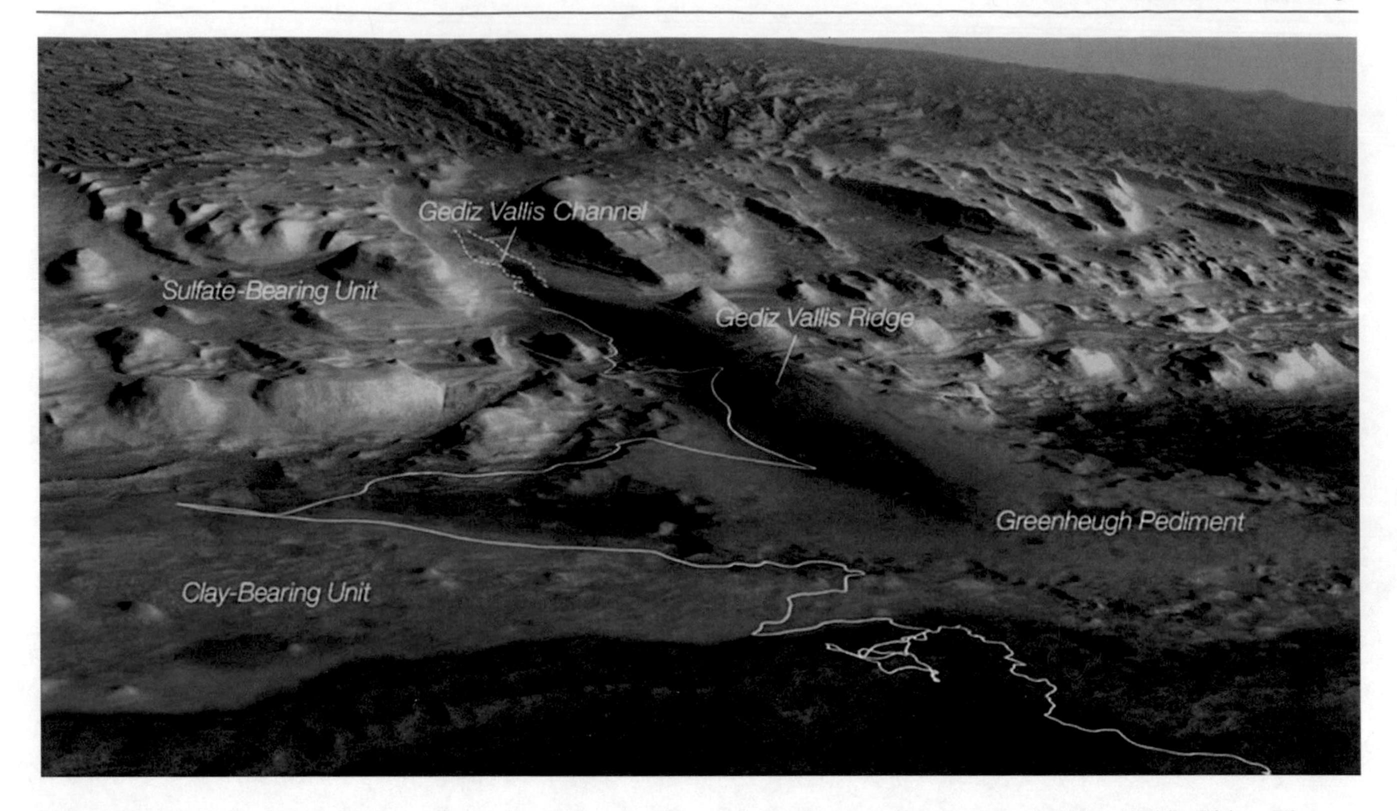

Fig. 9.30 Major hydrology features from the sulfate-bearing and the clay-bearing areas to the Vera Rubin Ridge. The text below discusses the effect the dynamic flow of water and rock had on the downstream Vera Rubin Ridge. The relation of the terrain in this image to the Clay Bearing unit will be discussed in the next chapter. (Image courtesy of NASA/JPL-Caltech/ESA/Univ. of Arizona/JHUAPL/MSSS/USGS Astrogeology Science Center.) This magnificent image has also been sourced elsewhere as the MSL Science Team

The dynamic effects on the middle elevations of Mount Sharp can be seen in Fig. 9.30. This figure will be referenced heavily in the next chapter as well as here: it is included in this chapter to support the interaction among the Vera Rubin Ridge, the Clay-Bearing unit, and the Sulfate-Bearing unit.

The sulfate-bearing slopes to the east of the region of Fig. 9.30 were fed by Hesperian floods. There must have been a raging river, saturated with calcium sulfate and minor components of the Sulfate-Bearing section flowing through the Clay-Bearing unit. For convenience, let us call that river the Gypsum River, since it probably carried calcium sulfate, commercially called gypsum if it is hydrated.

This Gypsum River was restrained by the Vera Rubin Ridge, holding its southern flank. At the western end of the ridge, depending on the topography there, it may have turned to the North, descending more directly to wherever the surface level was at the start of the flood-stage rains of the Hesperian era.

The Great Floods The scene of Fig. 9.30 is based on digital elevation maps from orbital photography. The colored and labeled units have been contributed by several geologists.

Examining the behavior of this very special region, together with the conventional understanding of the Hesperian period, suggests a violent scenario. As the flooded river continued to carry its burden of dissolved calcium sulfate and debris through the Clay-Bearing unit, a new event was beginning. The broadening Gediz Vallis (formally named in 2017) was eroding, concentrating more of its run-off to the Gediz Vallis channel, along a line of maximum decent. As the sulfate was being soaked it grew denser until great sections of it slipped downward, lubricated by the flowing water.

The initial slips dislodged other masses, until much of the Gediz Vallis, at least the banks of the Gediz Vallis Channel, was in motion. The landslide of calcium sulfate and other debris thundered into the flowing Gypsum River at right angles. Debris entangled with displaced calcium sulfate mostly dropped to the bed of the river, forming the Greenheugh Pediment. This structure is in layers, indicating a sequence of such events, perhaps associated with ebbs and flows of volcanic eruptions elsewhere on Mars. Overflow debris along the Gediz Vallis channel formed the Gediz Vallis Ridge, a natural levee.

As the landslides and flood impacted the river, the dynamics of water saturated with calcium sulfate and massive debris was thrown against the Vera Rubin Ridge, first collapsing its edge and then depositing much of its momentum on top of the ridge, washing across it and then down the other side. Such events could have been occurred more than once.

Such a series of dynamic attacks on the Vera Rubin Ridge may well have been the fortuitous cause of the ramps Curiosity used to ascend to the top of the ridge and then descend. Such convenient ramps are rare along the ridge, especially along its southern scarp. The two ramps are essentially down slope of the Gediz Vallis channel.

This momentous event involves more than the Vera Rubin Ridge, yet the ridge does play a principal role in the story. The story needs a great deal of further development. Parts of the story may well be provided by analysis of Curiosity's study of the clay-bearing area, still under way as this book is being written. Still more will be supplied in years to come as Curiosity ascends the ridge and the Gediz Vallis.

Dynamic simulation: There is ample opportunity to explore the quantitative assumptions of the above story by examining the available topography of the Gale crater, including Mount Sharp. Additional investigation could be carried out by computer simulations. The dynamic aspects of the flooding water and landslides from the Gediz Vallis, Gypsum River and Vera Rubin Ridge would in the future also benefit from further study of physical models as well. A hydrology lab could build a scale model of the Gediz Vallis, the surrounding Sulfate Bearing unit, the Clay-Bearing unit, Vera Rubin Ridge and the terrain beyond it to the North. The model of the Gediz Vallis should include loose debris. Two flows of water intersecting, one from the Gediz Vallis and the other from the trough of the Clay Bearing unit should have an interesting result. Getting the scaling right would be the challenge.

10 The Clay-Bearing Unit

The clay-bearing unit has been described in topographic and geological maps from orbit, supplemented with in situ photography taken by Curiosity during its ascent to the Vera Rubin Ridge. A very good view of its context with the Vera Rubin Ridge north of it and the Sulfate-Bearing unit south and above it can be seen in Fig. 9.30. The history of this extraordinary scene has been discussed in the results section of Chap. 9.

As this book is being written, Curiosity is still exploring this unit and its surroundings, so this chapter will be limited to the investigation of Glen Torridon, the Central Butte, Western Butte, and Tower Butte and the Greenheugh Pediment, relying primarily on raw images from Curiosity's cameras. In this pause in its exploration, it is taking a long drive around a desert of deep, rippled sand to its next major goal, the Gediz Vallis Channel in the Sulfate-Bearing unit.

Figure 9.30 in the previous chapter shows an oblique view of the clay-bearing unit from the southern edge of the Vera Rubin Ridge. The surface of the clay region runs uphill, at about a 10-degree slope, like the top of the Vera Rubin Ridge. At the northern face of Mount Sharp, the unit is bounded below by the scarp of the Vera Rubin Ridge and above by a sharp rise of the edge of the Sulfate-Bearing unit. The surface of the clay-bearing unit is called Glen Torridon after the narrow valley of the Torridon River of Scotland.

The Greenheugh Pediment overlies the Glen Torridon in the southern two-thirds of the clay-bearing unit. It is composed of material that has come down the Gediz Vallis (valley) from the Sulfate-Bearing unit. An interpretation of the role water played in the ancient history of Mars is described in the results section of Chap. 9. That story was derived from the orbital investigations that are summarized in Fig. 9.30. No doubt the interpretation will be expanded and revised in the future when Curiosity completes her mission of in situ investigation.

Curiosity left the Vera Rubin Bridge near the Lake Orcadie waypoint she had passed on sols 1977–1982 on the walk-about. There, part of the southern scarp of the ridge had collapsed, leaving a traversable path down to the clay-bearing unit. The area of the new sub-unit nearest the Vera Rubin Ridge has been named Glen Torridon, after a valley of a river in northwestern Scotland. The scene after descent, looking to the northeast, is seen in Fig. 10.1. It shows the southern scarp of the Vera Rubin Ridge, running to the East past the path Curiosity descended to explore the clay-bearing unit. The image in Fig. 10.1 shows the area of the highest concentration of clay, as measured from orbit. The foreground of this image shows rocks that appear similar to the Murray formation. After sampling similar rocks, Glen Torridon was confirmed to be a water-deposited mudstone, and is considered to be the Glen Torridon member of the Murray Formation.

C. J. Byrne, *Travels with Curiosity*, https://doi.org/10.1007/978-3-030-53805-7_10

Fig. 10.1 Curiosity took this dramatic picture after descending from the Vera Rubin Ridge to the clay-bearing unit, looking back at the ridge from the undulating bed of the ancient river. The southern scarp of the ridge looks much more eroded than the northern side. There are long talus slopes of fine debris running down to Glen Torridon. It is unlikely that much exposed contact will be found. The rock layers in the foreground look much like the Murray formation we have seen so far. Curiosity took this left Navcam image on sol 2316. (Image courtesy of NASA/JPL-Caltech)

Pebble Field Curiosity encountered a large field of pebbles while driving east along the upper scarp of the Vera Rubin Ridge (see Fig. 10.2). They could be concretions, washed down from the sulfate-bearing area, analogous to the "blueberries" found by Spirit and Opportunity. In any case, they seem to be concentrated, like pebbles on some beaches.

After finding an exposure of flat rocks suitable for drilling, Curiosity drilled and took samples from two adjacent rocks. See Fig. 10.3 for a selfie, the two drilled holes, and the context image showing a promontory of the Vera Rubin Ridge, the floor of the clay-bearing unit, and Mount Sharp on the horizon. A close-up of the two drilled holes is shown in Fig. 10.4.

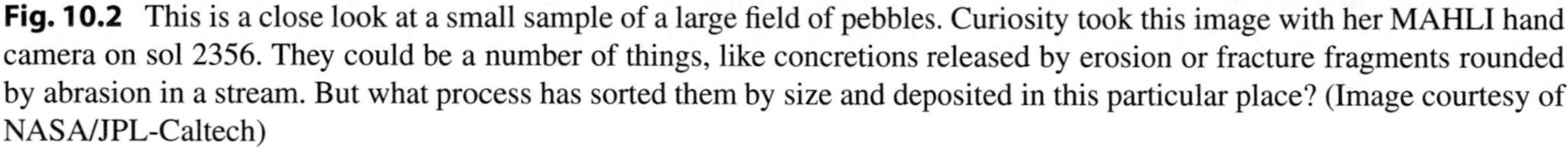

Fig. 10.2 This is a close look at a small sample of a large field of pebbles. Curiosity took this image with her MAHLI hand camera on sol 2356. They could be a number of things, like concretions released by erosion or fracture fragments rounded by abrasion in a stream. But what process has sorted them by size and deposited in this particular place? (Image courtesy of NASA/JPL-Caltech)

This site showed a maximum concentration of clay shown by the CRISM spectrographic camera in the Mars Reconnaissance Orbiter. Curiosity's SAM suite of instruments confirmed the high clay content with the highest value it had ever measured. However, other SAM results had been almost as high. Of course, orbital and in situ measurements must be understood in the context of averaging areas.

After drilling the two samples, Aberlady and Kilmarie. Curiosity continued her walk-about by crossing the clay region in the direction of the Greenheugh Pediment. The track map is shown in Fig. 10.3. As the map shows, Curiosity's path was relatively calm as long as it ran near the scarp of the Vera Rubin Ridge, but became rougher as it approached the pediment (Fig. 10.5).

Teal Ridge is a large outcrop of rock that has its laminated surface exposed (see Fig. 10.6). It marks a transition from the relatively smooth surface of Glen Torridon to the much rougher terrain as Curiosity approached the Greenheugh Pediment. It has been suggested that it may be an example of the Stimson formation at a higher elevation than where its usual occurrence.

The Central Butte massive outcrop (Fig. 10.9) may be related to Teal Ridge. Like the smaller outcrop, it presented an opportunity to explore the edges of laminations with the contact instruments. However, before she got to Central Butte, Curiosity paused to drill a couple of samples at Glen Etive (see Fig. 10.7).

Fig. 10.3 The rock slab just in front of the rover has the drilled hole called Aberlady near the bottom edge of the image. The rock above and to the left on the image has the drilled hole called Kilmarie. A promontory of the Vera Rubin Ridge, with its upper layers exposed and the lower layers shielded by a talus slope and bench, is in the upper left of the image. Mount Sharp is near the center top of the image. Curiosity took this image with its MAHLI camera on sol 2405. (Image courtesy of NASA/JPL/Caltech/MSSS)

The sampling site known as Glen Etive is near the edge of a plateau that appears to be at the very edge of the Greenheugh Pediment. It marks the end of one of the largest lahar-like flows of sediment and debris. As such it was selected for two drill samples. One is for the usual powder samples, and the other is for a rare experiment in wet chemistry, the second so far. The solvent chosen is hoped to allow the analysis in SAM to detect fragile segments of organic chemicals. The high concentration of clays here is hoped to stimulate production of early life on Mars or to protect complex organic structures, whether produced biologically or randomly by inorganic processes involving rocks and water, as methane can be assembled.

The second sample at Glen Etive 2 is especially meaningful because it is only the second sample to be tested with a stored wet reagent. This material reacts with organic molecules to enhance the dtection of complex organics such as RNA or DNA, or broken secments of such material. After taking the samples of Glen Etive 1 and Glen Etive 2, Curiosity proceded further on the Greenbaugh Pediment toward and up Western Butte (see Fig. 10.9).

The Central Butte: the map in Fig. 10.8 shows the route of a close examination of the Central Butte. The sites of Glen Etive 1 and 2 are just a short distance North of this map.

The next objective for Curiosity in this series of outcrops in the clay-bearing unit is the Western Butte. A long-range view of that feature is shown in Fig. 10.10.

As Curiosity approached the Western Butte (see Fig. 10.11), she climbed onto a new layer of Greenheaugh Pediment. The Murray Formation here is badly scarred, as if debris was dragged over it. That distinguishes the pediment itself from an outer layer that is deposited by the flooding water. Curiosity proceded to cimb the eastern side of the Western Butte (see Fig. 10.10)

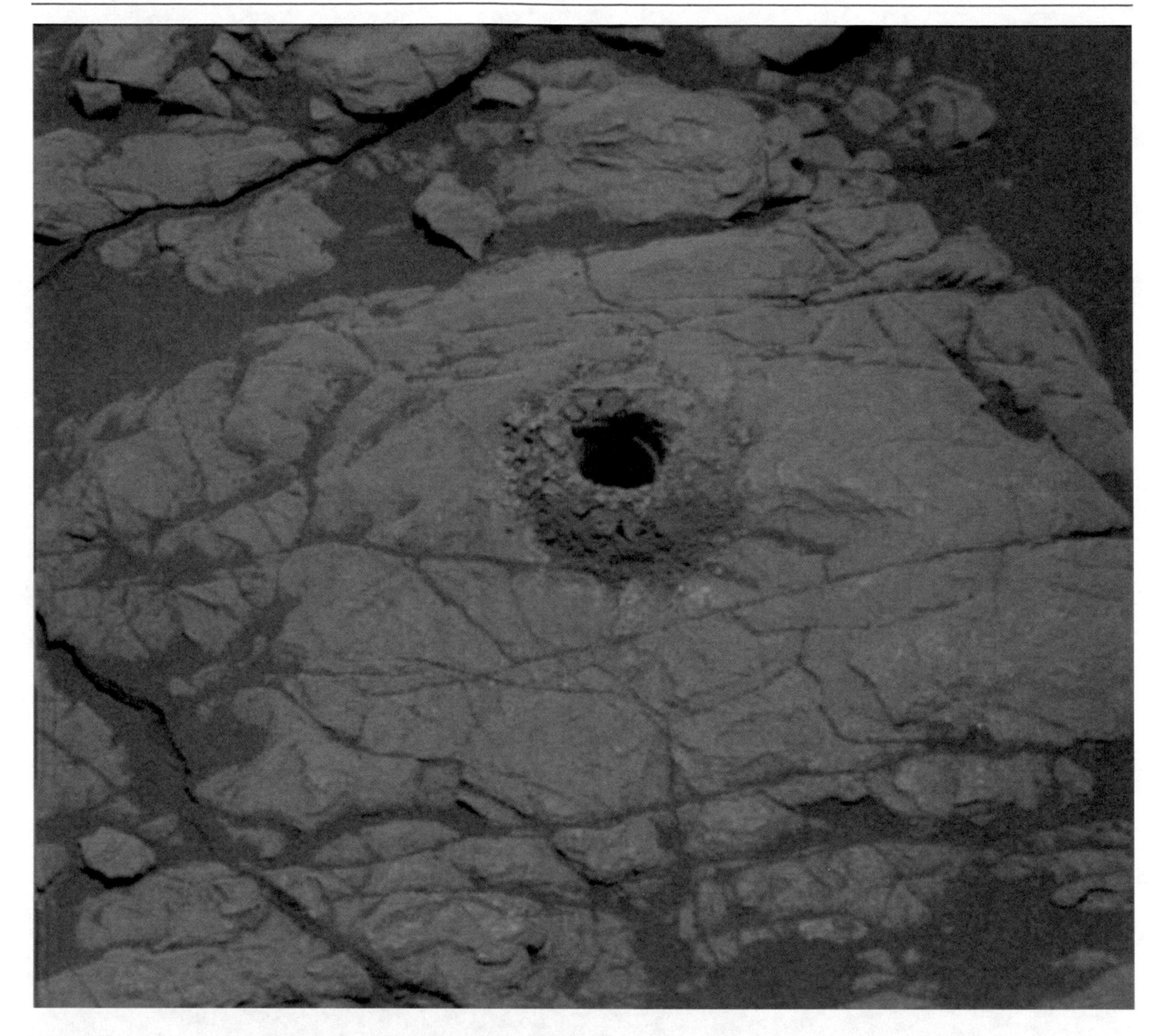

Fig. 10.4 This is the first drill hole, called Aberlady, and the first sample of the clay-bearing unit. It is the 21st successfully drilled sample that Curiosity has delivered from SAM for analysis. The tailings around the hole are interesting. They show the flakiness of the upper layers, which shattered as the drill penetrated. It is routine to examine the tailings of each hole with other contact instruments to compare their composition with that of the powdered sample delivered to SAM. (Image courtesy of NASA/JPL-Caltech)

The path of Curiosity between the samples of Aberlady and Kilmarie and Glen Etive 1 and 2 is shown in Fig. 10.5.

After taking the image of the Western Butte shown in Fig. 10.11, Curiosity moved forward and upward again (see Fig. 10.12) to examine the dark capping rock, possibly of the Stimson Formation. Individual rocks from the cap have tumbled down over the lower Murray Fomation rocks. Some of those rocks appear to have been pushed by wind, leaving tracks on the smooth light plates of the Murray Formation. It is possible that the Murray Formation rocks have unconformably covered dark rocks below. Note that as the elevation has increased at this face, the Murray Formation is no longer scarred by the heavy debris of the Greenheugh Pediment that was shown in Fig. 10.10, just scratched by float rocks (loose rocks).

The large laminated rock Teal Ridge in Fig. 10.6 was examined on sol 2440. It shows crossbedding in a bedrock layer and a contact over a rubbly layer.

Fig. 10.5 This shows Curiosity's path through the clay region up to sol 2568 (October 25, 2019). As usual when entering a new area, it did a walk-about to get the general nature of the area. This is not easy in this area because of the chaotic interaction between the Gypsum River that once carried runoff from the Sulfate-Bearing unit and the flooding from the Gediz Vallis coming in at a right angle (see Fig. 9.30). Between the sample Kilmarie, where Curiosity turned right to go south on sol 2506, and the sample Glen Etive 1 on sol 2523, Mars was in conjunction with the sun. Commands were inhibited for two weeks. (Image courtesy of NASA/JPL-Caltech)

Fig. 10.6 Teal Ridge, a large exposd rock, is an attractive opportunity for examining the edges of the laminations with contact instruments. This image is a mosaic of pictures Curiosity took with its MAHLI camera on sol 2440. (Image courtersy of NASA/JPL-Caltech/MSSS)

Fig. 10.7 Glen Etive1 and Glen Etive 2. These two drill holes allow comparison of the results of SAM analysis by the typical dry powder process and by wet chemistry. The dark streak near the center horizen is the southern scarp of the Vera Ruben Ridge, nearly a kilometer away. The eastern edge of Mount Sharp is at the upper right corner. Curiosity took the images for this mosaic with its MAHLI camera on sol 2553, October 11, 2019. (Image courtesy of NASA/JPL-Caltech/MSSS)

As Curiosity climbed higher, it went to the eastern of the Western Butte to stay on the smooth surface, which has been exposed there.

Curiosity drove a little higher to examine the capstore more closely (see Fig. 10.13).

Curiosity then drove around the the south where it could drive near the very top of Western Butte and look to the northwest along the edge of the lower unit of the Greenheau Pediment (see Fig. 10.14). The rover had been driving on that unit since drilling for the two Glen Etive samples.

Fig. 10.8 The yellow line is the track of Curiosity in the clay-bearing unit as she investigated the Central Butte at the far edge of the Greenheugh Pediment. She took the image of the Central Butte (Fig. 10.9). from the position labeled Sol 2565. (Image courtesy of NASA/JPL-Caltech, a part of the interactive map "Where is Curiosity?" for sol 2582)

The map in Fig. 10.8 shows a plan for a campaign to investigate the rocks at the foot of the North face of the Central Butte (Fig. 10.9).

Fig. 10.9 This very large exposure of the Central Butte may have been a relic of erosion of the bedrock of the Gypsum River. It has a streamlined shape, as if it were formed by the flow of water. However, it has strong crossbedding and may have been formed by sand dunes blown by wind in a dry period, like the examples of the Stimson formation lower on Mount Sharp. Curiosity took this image with her left Navcam on sol 2565 or her right Navcam on sol 2567, from the same position, to make a stereo frame. (Image courtesy of NASA/JPL-Caltech)

Fig. 10.10 The Western Butte in the near distance, past scarred Murray Formation rocks. Curiosity took this picture during the approach to the Western Butte (sol 2615 to 2626) with her right front Hazcam. Note that the Murray Formation extends near the top of the Western Butte on the eastern side. (Image courtesy of NASA/JPL-Caltech)

After investigating the Central Butte, Curiosity finally got to climb up the side of a butte, following the Murray formation unit that can be seen on the Western Butte in Fig. 10.10. The path can be seen up close in Fig. 10.11, Fig. 10.12, Fig. 10.13, and Fig. 10.14.

From its position high on Western Butte, Curiosity took a series of four pictures with its left Navcam on sol 2639 (see Fig. 10.15). They show the next level of the Greenheugh Pediment, a trough, and then the Sulfate-Bearing unit. The Gediz Vallis Channel runs down into the trough (see Fig. 10.16).

An enlargement of the rightmost image of Fig. 10.16 is shown in Fig. 10.17, showing detail of the mouth of the Gediz Vallis Channel.

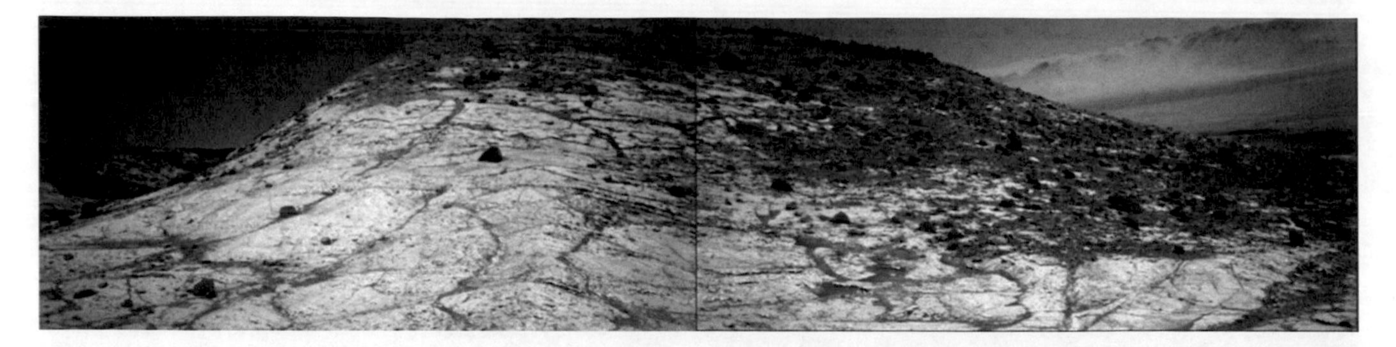

Fig. 10.11 Western Butte is one of several buttes in the clay-bearing unit, in this case also in the Greenheugh Pediment. There is a suggestion that these Buttes may have been capped by very high elevation examples of the Stimson Formation. If so, they might indicte the top of the Murray Formation, unless it continues under the Sulfate-Bearing unit. The two images for this mosaic were taken on sol 2632 by Curiosity's Right Navcam. (Image courtesy of NASA/JPL-Caltech/Charles Byrne)

Fig. 10.12 Nearing the top of the south face of the Western Butte, Curiosity took this image with her right Navcam on sol 2633. Note that some of the blocks of dark rock that have eroded from the cap unit have flat cleavage surfaces. (Image courtesy of NASA/JPL-Caltech)

Fig. 10.13 Curiosity came near the top of Western Butte to take a good look at the cap stone there on sol 2634 (right NavCam). (Image courtesy of NASA/JPL-Caltech)

Fig. 10.14 Curiosity moved south to where she could safely near the top of Western Butte and see the neighboring cliff on sol 2638 (right Navcam). (Image courtesy of NASA/JPL-Caltech)

During flood stages of the an ancient river in the Gediz Vallis, water flowing in the Gediz Vallis Channel would have run into the Gypsum River. To the rignt (west) of the channel is a thin dark stratum that was used to level the image. That stratum, that tops a laminated unit, may mark the surface prior to the formation of the Sulfate-Bearing Layer. The stratum continues to the left of the Gediz Vallis Channel. The planned ascent path is to climb in that channel to investigate the Gediz Ridge and the Sulfate-Bearing unit.

From its risky perch on the shoulder of the Western Butte, Curiosity looked out across the pediment toward the Sulfate-Bearing unit and the Giza Vallis Channel (Fig. 10.14)

Fig. 10.15 Curiosity took the images for this montage as it perched at a strong angle near the top of the Western Butte. An excellent oblique image of the geology of this area is shown in the previous chapter, in Fig. 9.30. (Images courtesy of NASA/JPL-Caltech. Assembly of the montage by Charles Byrne)

Fig. 10.16 This is the right-most image of Fig. 10.15, enlarged. The Gediz Vallis Channel is shown at the bottom of the Gediz Vallis (see Fig. 9.30 for an overview of the general area). The black area in the left center of this image is the caprock of a butte in the foreground that obscures the outlet of the Gediz Vallis Channel. (Image courtesy of NASA/JPL-Caltech. Rotation of the image to approximately level it by Charles Byrne)

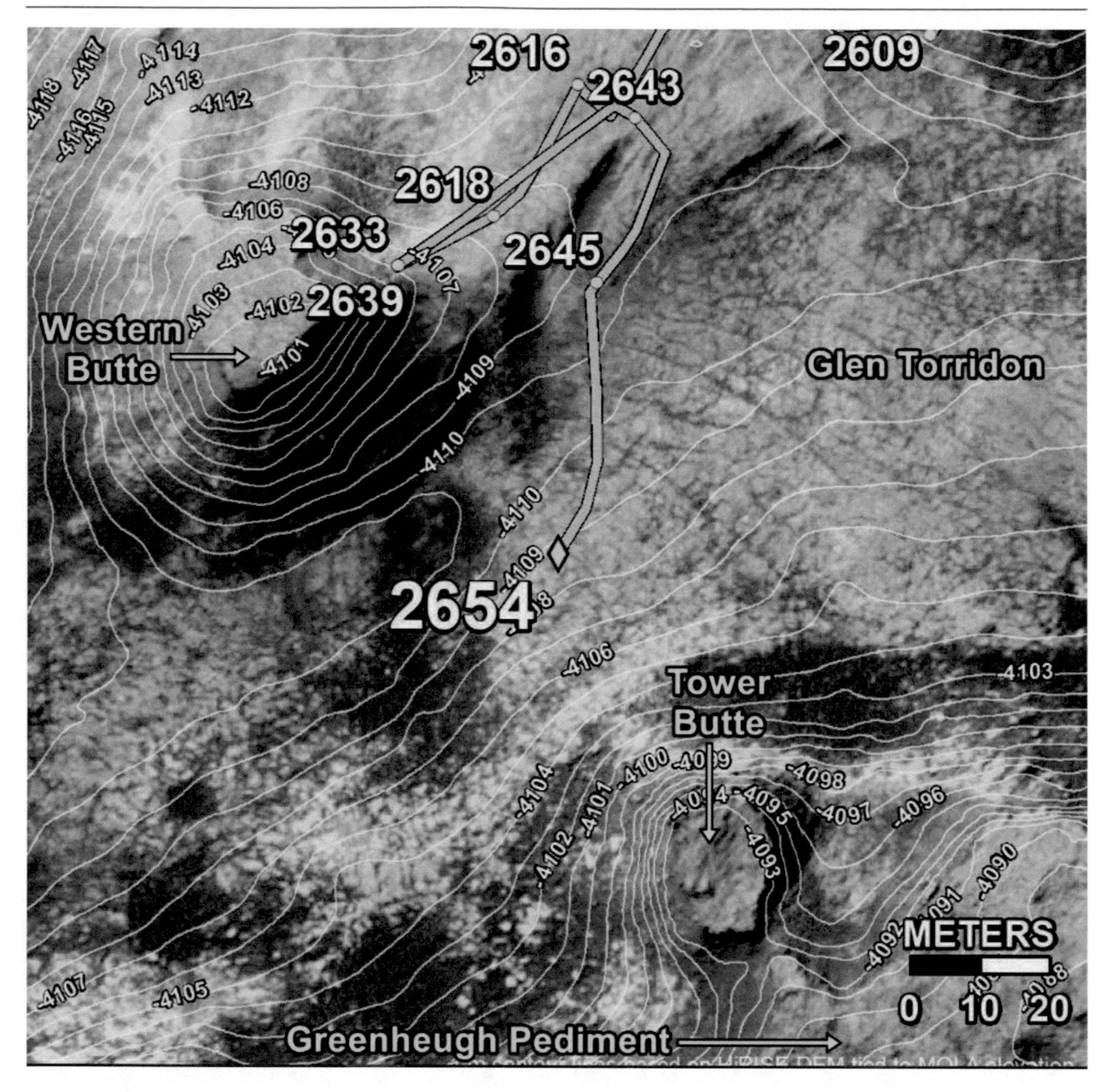

Fig. 10.17 Traverse from Western Butte toward Tower Butte from "Where is Curiosity" web page. (Image courtesy of NASA/JPL-Caltech)

The topographic map in Fig. 10.17 shows the path of Cutraveled near the top of the western Butte as she took pictures like a tourist and then rolled back down. She turned around and went back up, toward the pediment and the Tower Butte.

The large raised elevations on either side of the channel are the ends of two ridges that may have resulted from the deposit of sediment during flood stages of the Gediz Vallis Channel. The ridge west of the Gediz Channel is called the Gediz Vallis Ridge and may have received deposits from a part of the water from the Gediz Vallis River that did not flow down the Gediz Vallis Channel. Some of the dark shadows in the cliff may mark caves or perhaps the outflow of underground rivers.

After Ciuriosity left Wesern Butte 2639, it traversed in three drives for a total of 150 meters, taking about 15 sols, to a waystop at 2654, midway between Western Butte and Tower Butte (see Fig. 10.17).

Fig. 10.18 As Curiosity descended down a fortuitous ramp from Western Butte toward Tower Butte, she came across this feature called Balgy, filled with sand, that might once have been a stream. (Image courtesy of NASA/JPL-Caltech)

As Curiosity descended from Western Butte and turned toward the Tower Butte, she encountered a lineal sandy patch (see Fig. 10.18). The feature, which was called Balgy, can also be seen in Fig. 10.17. It was named after the River Balgy in Scotland. Balgy on Mars was seen as mysterious, but at least one similar feature has been seen before, in Fig. 9.5. That feature, seen on the ascent to the Vera Rubin Ridge, was a similar straight-sided ditch now filled with sand, seeming to have been cut by flowing water.

The two buttes, when they were partly eroded, might have constricted a flow of water running down from the Greenheugh Pediment, forming a valley in the pass. A linear vertical fracture in the base rock might have expedited the process of erosion.

Fig. 10.19 Map of Curiosity's path to take a Mastcam mosaic of Tower Butte on sol 2658. The base map is based on images from orbit. The figure was taken from the web page "Curiosity Rover current track map". The track also shows the path Curiosity took to suvey the rockey edge of the Greenheugh Pediment and finally reach the pediment. (Orbital image and track courtesy of NASA/JPL-Caltech)

Fig. 10.20 Mosaic of the north face of Tower Butte trom images of Curiosity's Mastcam 0n sol 2658. (Image Courtesy of NASA/JPL-Caltech/MSSS)

Driving closer to the Tower Butte (see path on Fig. 10.19), Curiosity took images for a mosaic of Tower Butte (Fig. 10.20) on sol 2054.

The background orbital photograph of this map (Fig. 10.19) from the "Curiosity current track map" web page is different from the toptographic one from the "Where is Curiosity?" track map (Fig. 10.17). The track map of Fig. 10.19 is usually available a few days earlier because it relies on the internal guidance of the rover, while the track on the Topographic map is corrected from visual clues from the cameras.

Fig. 10.21 The west face of the Tower Butte. Curiosity took this picture with her right Navcam on sol 2658. (Image courtesy of NASA/JPL-Caltech)

Fig. 10.22 On Vallantine's day, sol 2674, Curiosity sent down a MAHLI picture of a heart-shaped sunbeam shining on Mars rock. (Image courtesy of NASA/JPL-Caltech/MSSS)

After taking the mosaic of Tower Butte, Curiosity drove to the butte's west side for a close-up (see track on Fig. 10.19 and image of the west face of Tower Butte on Fig. 10.21).

Possibly because of the talus slope and angular rocks shown in Fig. 10.21, it was decided not to climb it like had been done on Western Butte. Curiosity's attention was turned) to the edge of the Greenheugh Pediment, after drilling one more sample before that investigation. A thorough process of preparing to drill was begun; selectiion of bedrock, brushing, examination, minidrill, etc.

During the drill preparation sequence, the MSL Science Team Members needed a tension break. Fortunately, it was sol 2674, Valentine's Day, 2020. Perfect. See Fig. 10.22.

Dawn Sumner, a planetary scientist from the University of Utah wrote the next mission update blog, titled "Sols 2676-2679: 4 Sols of Love for Curiosity!'. She also wrote a poem (see box)

Sols 2676-2679: 4 Sols of Love for Curiosity

Drilling on Mars! It's so super cool! SAM bakes up our sample Using many a joule

Dawn Sumner

Fig. 10.23 *Hutton* is the last sample (#24) before the Greenheugh Pediment. It was taken from a slab of bed rock that looks like it might be Murray Formation. SAM will analyze it and tell us more. (image courtesy of NASA/JPL-Caltech). Curiosity took this picture on sol 2687 with the MAHLI camera. (Image courtesy of NASA/JPL-Caltech/MSSS)

*The Hutton Drill Sample***,** after successful drilling, examination, and transfer to SAM, is shown in Fig. 10.23.

The map in Fig. 10.24 shows that the Greemheugh Pediment was deposited in layers. This indicates that there were a series of floods or landslides carrying debris into the Clay-Bearing unit.

Following the successful collection of the Hutton sample, Curiosity turned her attention to the examination of the edge of the Greeneugh Pediment, a meter-high wall of loose rock, not safe to climb directly. A large scale view of the context of the pediment is shown in Figs. 10.24 and 10.25.

Fig. 10.24 This picture from the Mars Reconnaissance orbiter shows that there have been two layers of debris from the Sulfate-Bearing unit deposited on the clay-bearing unit. Both layers are runoff from the Gediz Vallis Channal and other parallel sources in the Sulfate-Bearing unit. They share patterns of numerous craters, probably from a cluster of meteors but possibly from drainage into voids in the debris fields (sinkholes). The material in the debris is likely to contain calcium sulfate, which is water-soluble. The scale key in the lower left corner is 400 m long. (Image courtesy of NASA/JPL-Caltech/MRO)

Fig. 10.25 This image taken with the Front Hazcam on sol 2692 is one of many pictures of the edge of the Greenheugh Pediment that were taken to find a path forward. The rock wall would have to be climbed if Curiosity would be able to take a short-cut to the Sulfate-Bearing unit. The most promising path through is just left of center of this image but it is too narrow to get through easily. (Image courtesy NASA/JPL-Caltech)

The Greenheugh Pediment and the Rock Wall

The boundary between the Greenheugh Pediment and the earlier layer that runs beyond it to the north is edged with a wall of loose rock about a meter high, running nearly continuously, at least in the area where Curiosity has approached it. It has the nature that is analagous to that of a glacial moraine. The earlier shallow layer may have had a greater ratio of water to rock and the later layer more rock than water. The first layer may have been more like a flood carrying debris and the second more like a landslide.

The rock wall is visible in Fig. 10.25 between the earlier shallow pediment and the thick later pediment. It may have been formed by an interaction between the two layers as the later layer came to a halt.

Which Path to Take?

The Rock wall on the edge of Greenheugh Pediment has been a subject of concern from sol 2659 (January 28, 2020) to sol 2647 (February 14, 2020). Curiosity took many pictures of a rock wall about 1 m high that obstructed her path to climb up onto the Greenheugh Pediment.

Planned Path There were two possible paths to get to the Sulfate-Bearing unit, their next goal beyond the pediment. The planned path, involved a long traverse eastward, a short drive northward to get around a large sand dune area, and a long traverse westward to get to a short drive on the pediment. Then another short drive would bring Curiosity to the Gediz Vallis channel in the sulfate-bearing unit.

Alternative Path The new idea was to get up on the pediment near the Tower Butte and cross the pediment (about a kilometer or two) and then go directly to the Gediz Vallis channel.The question took no time at all to escalate to the MSL Project Group, who brought Paul Grozinger back to help the difficult decision.

Two weeks passed as they moved Curiosity back and forth looking for a good path to climb the loose rock wall to get up and over to the pediment. The edge of the pediment looked like Fig. 10.25 (or worse), in all directions but back (leading to the planned path).

Finally, the drivers of Curiosity felt they had a route over the wall that looked difficult but possible. It starts in the narow opening indicated in the legend of Fig. 10.25, and involves two sharp turns (each involving steep banks) to reach a promising path. The start is shown in Fig. 10.26, and the finish in Fig. 10.27.

To quote Michelle Minitti, Planetary Geologist at Framework, who wrote a mission update blog,

> "Kudos to our rover drivers for making it up the steep, sandy slope below the Greenheugh pediment … and delivering us to a stretch of geology we had our eyes on even before we landed in Gale crater!"
>
> Michelle continued: The geology planning group honored the achievement of making it here by getting our cameras and laser on every little bit of rock we could manage. MAHLI and APXS will analyze Galloway Hills cleared of dust before and by the DRT, and Ardwell Bay. "The former is on a smoother, flatter part of the sandstone we are parked on, and the latter is an example of the resistant features that dot the sandstone in this part of the pediment. MAHLI will also acquire a mosaic looking at a package of sandstone layers at the bedrock target Chinglebraes."

The significance of sandstone is that the Murray formation sediment, deposited at a lakeshore, had a composition dominated by mudstone. The Stimson formation was deposited as silica sand and formed into dunes by dry wind. Neither formation was pure in composition: instead, the characterization of mudstone or sandstone is determined by dominance.

Fig. 10.26 This is the start of the pass to the pediment, as shown in the context of the rock wall in Fig. 10.25. Curiosity will drive to the right wall of the pass, run up the rocks slowly, taking a steep bank, and turn left. It will then go down the rocks and across the narrow pass, go up the rocks on the left side, turn right, go to the edge of a crater and stop there for photography (see Fig. 10.27). The path is shown on the track map of Fig. 10.19. (Image courtesy of NASA/JPL-Caltech)

Geologists, with help from other disciplines, will decide whether the Greenheugh Pediment unit will be an existing formation or a new formation (or combination of formations).

Curiosity continued, turning right to move away from the crater and proceeded southwest on a low slope. On sol 2700 she paused at a low barrier to take a large Mastcam mosaic of the scene, including its future path, the Gediz Vallis Channel in the Sulfate-Bearing unit and the intervening Greenheugh Pediment that Curiosity hoped to conquer.

Fig. 10.27 Finish line! This looks more risky than it is. The track map shows that Curiosity stopped a few meters short of the big black block and then bumped forward to this position, the finish, to the pediment. It was certainly courteos of some Martian to leave that old-style engine block in place, just in case. The lighter-toned rock fragments which appear to have been blown out of the crater are showing classic shock patterns: cones with points in the direction of the impact point. The picture was taken by Curiosity's Right Navcam on sol 2692. (Image courtesy of NASA/JPL-Caltech)

Mastcam images were taken from Curiosity's position in Fig. 10.28 for a large mosaic, covering a greater area at highter resolution than this beautiful Navcam image. Then Curiosity easily overcame the low barrier and went looking for a nice smooth sandstone bed rock to drill. Yum! See Figs. 10.29 and 10.30 for a tasty treat.

Fig. 10.28 Moving over the ejecta field from the crater to her left, Curiosity encountered a low barrier on sol 2700. The very light block of rock in front of her, with smoothed edges, may be from the Sulfate- Bearing unit where the landslide started. Top center of this image, in dark gray, is the outlet of the Gediz Vallis Channel. The entire area in between Curiosity and the Channel is the Greenheugh Pediment. The mountain in the upper right is the Gediz Vallis Ridge. (Image courtesy of NASA/ JPL-Caltech)

Just beyond the rockwall in Fig. 10.28 , a candidate rock for a pediment drill sample was found (see Fig. 10.29) It was named "Edinburgh.

Fig. 10.29 This is the discovery image of the rock Edinburgh (bottom, just left of center) showing its location near much lighter rock and the dcge of a crater. Curiosity took this image with her right Navcam on sol 2700. (Image corurtesy of NASA/JPL-Caltech)

Fig. 10.30 The Edinburgh rock selected and brushed for sampling. Notice the strong color difference between the gray rock itself and the thick dust coating, the equivalent of "desert varnish" in the American southwest. (Image corurtesy of NASA/JPL-Caltech)

The rock slab Edinburgh survived examination by brushing and APXS investigation to supply sample #25. Its portrait, after brushing its dusty coating away, is in Figs. 10.30 and 10.31.

U-Turn! After a large Mastcam mosaic was examined, the conclusion was that the Greenheugh Pedment was so rough and broken up with holes, some by impact and some possibly by sinkholes, that it would be better to turn around and take the previously planned path forward. The MSL Science Team has a rich stock of information about the geology of the pediment now and Curiosity is likely to have an opportunity to visit it again as she approaches the Gediz Vallis Channel.

Nodules On the way back, she stopped just short of the rock wall to examine, with the XSPS instrument, some nodules that were laying on the bedrock. The word nodule is derived from the Latin for knot. In geology, a nodule is a small rock that seems to be out of place. In this context, the nodules might have come from a deep layer below the pediment, ejected to its rim by the meteorite impact that caused the crater east of the path Curiosity took after she went through the rock wall. We know it was an impact crater, not a sinkhole, by the shock cones laying on the rim (see Fig. 10.27 and its legend).

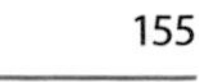

Fig. 10.31 Sample Edinburgh taken on sol 2711 by Curiosity's right Navcam. (Image courtesy of NASA/JPL-Caltech)

Results

In Situ Confirmation of Clay The data gathered by Curiosity agrees with the orbital data prediction that the clay-bearing unit has high levels of clay in its composition and is not only worthy of exploration but also has a strange beauty reminiscent of heavy rapids on Earth. has confirmed that the area on Mars it's exploring, called the clay-bearing unit, is well deserving of its name. Samples the rover drilled at rock targets called Aberlady and Kilmarie in the Torridon region, seen to be high in clay from orbit, have revealed the highest concentrations of clay minerals found so far. The drill hole for Aberlady is shown in Fig. 10.4.

This narrow band of clay-dominant material, seen from orbit before the target selection for the MSL, was one of the most attractive attributes of Gale crater. Clay is an extraordinary crystalline mineral: it forms in 2-dimensional sheets that alternate in their structure. Between those sheets are an array of molecules, often including water. The water is what makes clay moldable and also suggests these layers of clay, which are potentially habitable with the right minority minerals, as templates for the origin of life. Early in the exploration of the clay-bearing layer, the science instrument CheMin, which receives and analyzes samples before the more time-consuming full analysis of the thermal spectrometer part of SAM suite, detected a rich clay component, soon confirmed with more detail by the additional data of a full SAM analysis.

These two samples, taken from the part of the clay-bearing unit that is nearest the southern scarp of the Vera Rubin Ridge, were also much lower in hematite than the VRR, unlike the northern scarp, which had similar levels. The northern scarp shows much more typical erosion, with a full talus slope at its base.

Waves of Clay The visual images of the floor of the clay-bearing unit, especially south (uphill) of the early walk-about track, going east along the Vera Rubin Ridge, away from the Greenbaugh Pediment, show wave-like patterns, which could either be linear dunes, perhaps lithified. Alternately, the waves could be induced in soft clay layers of a river bottom. Also, closer to the walk-about track, there are rocky outcrops similar to features called danangs, that often form in flood stages of flowing water.

Field of Round Pebbles Another remarkable feature of the Clay Bearing unit are the concentration of rounded pebbles (Fig. 10.2), not seen in any other unit yet. They could have come down from the Sulfate-Bearing unit, if they formed there in hollows of the sulfate. Or they could have formed where they are, from layers of minerals laid down in layers like those of clamshells. Such rocks are called oolites because they look like eggs. The mineral might have come from saturation or suspension in water; if so, they could be made of calcium sulfate or clay.

Low Bedrock Density While Curiosity was in the Clay Bearing unit, a paper by Kevin W. Lewis et al., “A Surface Gravity traverse on Mars indicates low bedrock density at Gale crater” was published in Science on February 1, 2019. The study was based on data from engineering accelerometers used to determine attitude of the rover. Based on data that Curiosity had been taking as it rose through the lower layers of Mount Sharp, the investigators were able to measure the change in gravity with elevation, and infer the density of the surrounding rock. The result was 1.6 gm/cm^3, much lower than expected. For example, the density of sandstone is 2.65 gm/cm^3. One possibility is that the rock of Mount Sharp is porous. Another is that there is a hollow volcanic pipe or other large voids (caves?), within the mountain. In any case, previous suggestions of sediment completely filling Gale crater would result in a higher density than was observed.

Methane While in the Clay Bearing unit, specifically the Glen Torridon area, in sight of a striking scarp (Figs. 10.32, 10.33, and 10.34) Curiosity measured the largest burst of methane in its 7 years of observation on Wednesday, June 21, reported to Earth on Thursday, June 22, in time to plan a new observation on the following weekend. On June 21, the methane concentration at Curiosity was measured, by SAM as 21 ppb (parts per billion by volume). On June 24, the concentration was down to the background level. Experimental error? Not likely because a search for methane events in orbiter data found a significant one on June 21 in Gale’s vicinity. The center of the orbiter’s event was northeast of Gale, in an area called Aeolus Mensae. A mensa is like a mesa. This is an important contribution: independent observations of a methane plume and its puzzling brief presence in the local atmosphere.

Fig. 10.32 After backing out of the top layer of the Greenheugh Pediment and going back through the narrow pass through the rock wall, Curiosity had successfully retreated from the thick rocky layer. She went down the steep slope she had climbed on the way up to the rock wall and turned East to follow the original plan. Then, Curiosity took the image of the north face of the Tower Butte with her right Navcam

Fig. 10.33 On sol 2442, June 19, 2019 when Curiosity detected a large plume of methane amounting to 21 ppb (parts per billion), she took this image with her right Navcam. The location is north of the Central Butte. This observation was confirmed by orbital detection of methane in the atmosphere, not far away. The consistent pair of observations, together with many individual sporadic observations from Earth as well as by Mars missions, convincen planetary scientists that there was a persistant (although sporadic) release of methane at Mars. (Image courtesy of NASA/JPL-Caltech)

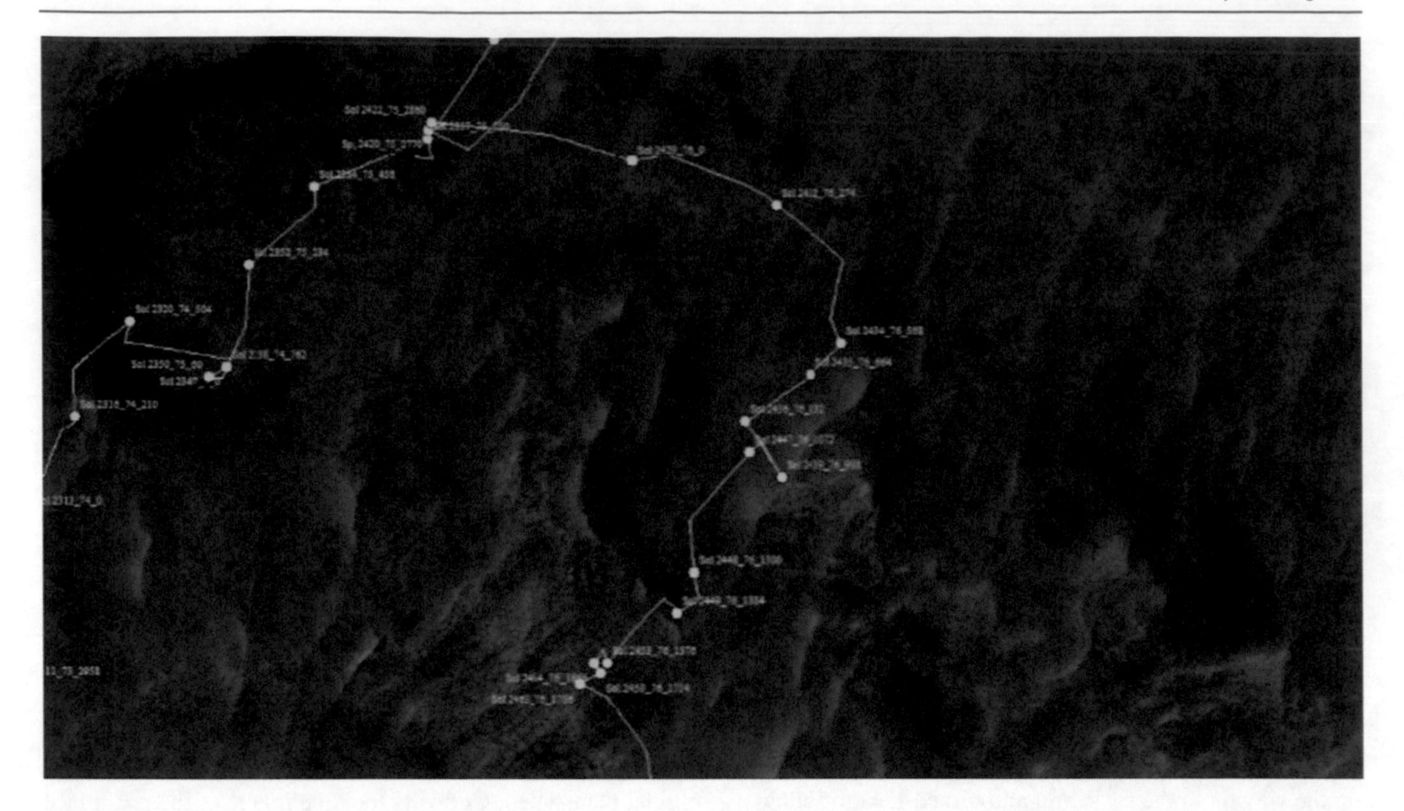

Fig. 10.34 This yellow line is the track of Curiosity in the clay-bearing unit. The Glen Torridon region of the Clay unit is at the upper left, with the edge of the Vera Rubin Ridge in the corner of the image. Curiosity left Glen Torridon to cross waves of either sand or river bottom covered with sand, continuing her walk-about. At sol 2442, when she measured the plume of methane, she was near the lower center of this image. The base image is orbital photography. This is a screen capture from the interactive NASA web page "Where is Curiosity". (Image courtesy of NASA/JPL-Caltech)

Dr. Aswan Vasavada, MML Project Scientist, JPL
"The methane mystery continues. We're more motivated than ever to keep measuring and put our brains together to figure out how methane behaves in the Martian atmosphere."

Wet Chemistry Experiment At Glen Etive 2 (see the selfie in Fig. 10.8), the second wet chemistry analysis was done by SAM. The first one was done on a sand scoop sample at Ogonquit Beach to test the process because the drill had just been broken and it was feared that there would be no opportunity to get another powdered core sample.

The sample was taken in the clay-bearing unit, which was of interest because the structure of clay molecules is thought to provide a good framework for primitive life or for pre-life organic molecules to build. The sample was chosen in the shallowest layer of the Greenheugh Pediment, instead of the more clay-rich Glen Torridon, possibly for the higher level of sulfate available there, which tends to help preserve organic molecules. The results of the wet chemistry experiment are not yet available.

"We've been eager to find an area that would be compelling enough to do wet chemistry," said SAM Principal Investigator Paul Mahaffy of NASA's Goddard Space Flight Center in Greenbelt, Maryland. "Now that we're in the clay-bearing unit, we've finally got it."

Buttes, Mesas, and the Stimson formation The Stimson formation was first seen, but not fully understood at the Kimberly location. In the Pahrup Hills, it was seen in the Whale Rock and Telegraph Peak, but still not understood. By the time the Marias Pass observations were available and the geologists were confronted by massive mountains like Mount Stimson and the active dunes of the Bagnold Dune campaign, opinions converged (see box). The geologists came to accept that the difference between the Stimson formation's sandstone units was sufficiently different from the Murray formation mudstone units that separate formations were required.

Curiosity Update: Recapping the Bagnold Dune Investigation

Dr. Catherine O'Connell-Cooper, University of New BrunswickASU Red Planet Report, October 22, 2018, posted by Robert Burnham

As Curiosity continues on her journey up Mount Sharp …rocks we encounter contain evidence for changing environmental conditions. The fine-grained mudstones of the Murray formation show us that lakes were present in the past, whilst the sandstones of the Stimson are evidence for ancient dune fields.

As Curiosity was directed to higher elevations, the massive dominance of the Stimson foundation was less seen and the opinion grew that the Stimson formation was limited to lower elevations but the Murray formation continued at higher elevations. In recent times, in a paper published in 2019, it was suggested that the relatively rare mesas and buttes like the three that were investigated in the clay-bearing unit (Central, Western, and Tower Buttes) and others that have been seen remotely in the Greenheugh Pediment were formed as part of either a widespread Stimson formation that has been widely eroded or, alternately, only rarely lithified and then eroded easily. In other words, it may be that the sediment was deposited widely but only locally lithified.

Therefore, in the region that Curiosity explored, there was a range of elevations for the Stimson formation to be a dominant layer, covering the Murray layer widely, yet still leaving enough of the Murray formation so Curiosity could drive efficiently. That allowed her to avoid the Stimson formation or deep sand. One wonders if that is unique to the north face of Mount Sharp or is it general to other directions. Of course, that is also a tribute to the intense examination of orbital photography and topologic maps, and the planning of the Site Selection Navigation Group to assure that there is not only a promising path planned but also enough alternate paths to take if a problem was revealed on the ground.

At this point, the planned route still looks good to get Curiosity to the last goal unit (near the north edge of the Sulfur-Bearing unit). The shorter route that was not selected but is much shorter, turned out not the way to go. Still, the MSL Science Team did get to visit and sample the deepest layer of the Greenheugh Pediment. To the credit of the operations team, and especially the drivers, we got science done there. Now to drive the long traverse around a deep linear sand dune to investigate the Sulfate-Bearing unit.

11 Management, Performance, and Productivity of the MSL Science Team

Introduction

Carrying out a complex planetary mission takes a lot of planning and management to achieve a successful mission. That is particularly so for the Mars Science Laboratory Mission, which attracted the creativity of a large national, international, and interdisciplinary group of scientists willing to devote a significant part of their careers to the joint effort over an extended period of time.

This chapter departs from the path of exploration of Gale crater by Curiosity to examine the planning of the science operations of Curiosity and her instruments, as well as the changes in plans made in the course of her operations. The planning to be analyzed in this chapter was recorded in an open-source peer-reviewed paper (and its supplement) published in Science Reports on July 25, 20012, two weeks before the Curiosity rover landed at Gale crater on Mars. (see box).

■
Original planning documents summarized in this chapter:

Mars Science Laboratory Mission and Science Investigation by John P. Grotzinger et al.
and the accompanying supplement:

MSL Science Team Rules of the Road
Text is inserted in the summaries by this book's author:

Comments on applications of the plan and

Changes in the plan based on operational experience

The MSL Science Team is the group of scientists who will make the decisions to operate Curiosity and its tools to maximize the scientific results of the mission. They do this in cooperation with the JPL engineers who designed and tested Curiosity. The members of the MSL Science Team were mostly nominated by the Principle Scientists of the science instruments. Altogether, there are about 400 MSL Science Team members and Science Collaborators listed in Appendix: The MSL Science Team.

The Operations Plan Of the two documents listed in the box above, the first document is a training course text to explain what the science team needs to know about the Curiosity rover to participate in day-to-day planning groups. Formal guide lines for the management of the MSL Science Team during operations on the surface of Mars are in the Rules of the Road supplement. The MSL Science team is

C. J. Byrne, *Travels with Curiosity*, https://doi.org/10.1007/978-3-030-53805-7_11

also responsible for the release of data, interpretation of results, and communication of the results to the public and to the scientific community. The nearly 400 papers in peer-reviewed journals is an indicator of the magnificent performance of Curiosity, the MSL Science Team, and the engineers who worked to meet the challenges of the Mars surface environment.

The initial plans were based on the very successful operations of the Mars Exploration Rovers, Spirit and Opportunity and other earlier missions. After the first months of surface operations, the MSL Science Team realized that Curiosity, with its ten science instruments, was much more complex than the Mars Exploration Rovers. Changes were made in the methods of operations to increase the efficiency of making decisions. In part this was done to relieve stress on the Mars Science Team and in part to increase the efficiency of planning operations that take place over multiple sols associated with time and power intensive operations such as drilling and analysis of samples.

Summary, Comments, and Changes Comments on the success of specific parts of the plan and also on changes to the plan that needed to be made are inserted in the summaries of the planning paper and its supplement. To differentiate between the original documents and the comments and changes, the original documents are in italics, while this author's comments and description of operational changes made in the course of the mission are in normal print. Additionally, to highlight transitions, a passage from the original documents will be introduced by the source in brackets: either **[MSL Mission and Science Investigation]** or **[MSL Rules of the Road]**. The author's comments, summaries, and descriptions of operational changes will be introduced by **[CB Summary]**, **[CB Comment]**, or **[CB Change]**. The major changes are for the addition of an intermediate planning step, the Supratactical Process after early experience with surface operations. The process is described by "The MSL Supractactical Process" by Debarati Chattopadhyay et al. and a passage from the text will be introduced by **[MSL Supratactical Process]**. Finally, text taken from the Science Corner of the MSL mission's webpage is introduced by **[MSL Science Corner]**.

Summary of "MSL Mission and Science Investigation"

[CB Summary] The abstract of this tutorial paper gives an overview of the MSL, its rover, and its instruments. The body of the paper begins by summarizing the literature from 2000 to 2012, a period of orbital survey, describing the surface of Mars and the evidence for extensive flow of water in ancient times. Previous missions have established that while water in the form of ice exists near the poles, conditions today do not permit flowing water on the surface.

Yet the surface of Mars has been transformed by interactions with water through much of its history. That point had been made before Curiosity was launched. Curiosity and the MSL Science Team were to investigate how long flowing water and lakes have existed on Mars and what can be learned from examining the extensive sedimentary rocks with the complex array of instruments and tools. The MSL Science Team plans the route and actions of Curiosity and then consider the implications of the observations to learn about habitability, preservation (of signs of life), and variations in the environment of Mars.

Water in the Martian Past
[MSL Mission and Science Investigation]
The summary conclusion from all these discoveries is that the surface of Mars has been transformed by interactions with water throughout its history.

This is exciting for science, but also reassuring to the Mars Program officials who adopted the "follow-the-water" strategy. This strategy has worked well and the planetary science community is richer for it.

What's next?

Habitability and Preservation

[CB Summary] Two specific objectives for Curiosity's instruments and the MSL Science Team were set by the pre-landing document: to determine the habitability of the environment for lifeforms in the past if not the present, and to investigate conditions that would (or would not) permit the preservation of evidence of such life if it had existed.

The necessary conditions for a habitable environment were set as the presence of liquid water, a source of carbon, and a source of energy to enable organic metabolisms.

Hazards to the preservation of *biosignatures* (an object, substance, or pattern requiring a life form to create) were also discussed. One difficulty is ambiguity: to be convincing, a biosignature must be distinct from any physical process. Also, complex structures of organic molecules can be degraded by radiation and particles of the solar wind, or by oxidizing environments. Finally, fossil structures formed by encasing or substitution of minerals in organic structures can be distorted or destroyed by changing environments.

[MSL Mission and Science Investigation] *Thus MSL will be faced with a major challenge: both modern weathering processes (including radiation damage) and ancient diagenetic processes could conspire to inhibit the preservation of organic matter.*

[CB Comment] The discussion in the pre-landing document on habitability was partly to define what habitability means and partly to manage expectations about instrumentation to detect past life. Shortly after the successful landing, besides confirming the previous understanding of water on Mars described above, Curiosity's exploration provided an answer to "What's next?" Early on, evidence was found of ancient water being neutral in Ph, with the necessary evidence for elemental and mineral content needed to support life as we know it. As the exploration went on, it was understood that the view of "early Mars wet, later Mars dry" was too simple, and that there had to be a lengthy succession of wet and dry periods to establish the observed rock strata in Gale crater. To return to the pre-launch document's expectation of what's next, the concept of habitability of the environment was described as an objective for Curiosity and the MSL Science Team.

Many observations were made of ambiguous processes that are common in the presence of water, especially flowing or groundwater and are prolific at lower Mount Sharp. Crystal patterns in evaporate minerals have been seen that are similar to those left by bacterial colonies, for example. Positive evidence for the survival of somewhat complex hydrocarbons shielded from the solar wind by only a few millimeters of surface has been found, but a similar level of complexity has been found in interactions between rocks and water.

Studies of ancient Earth rocks tell us about the early evolution of the Earth's surface and atmosphere, which not only led to the appearance of primitive life but also to the interactive relation between the biosphere, the atmosphere, and the surface of Earth. Even if a diligent search for signs of past life on Mars is unsuccessful, we will learn how a lifeless rocky planet develops and learn by comparison with Earths history.

Water, even flowing water, is not sufficient for life; by observation, we have learned that even *extremophiles*—species that survive and propagate in conditions that provide minimal support, including unusually hot or cold temperatures—require a common array of elements, a supply of energy, and a limited range of temperature and a limited range of Ph, a measure of acidic and basic water. Even the extremophile species, which can survive even in the pores of rocks, have their limits. They all are built partly with organic molecules, hydrocarbons. So the goal is to find evidence that there was (or prove that there was not) a condition where life could have been present. As it turned out, Curiosity's data showed that the evidence for the ancient environment when the examined rocks had been formed was within the range of parameters that could support the life we know on Earth.

Even on the crater floor, Curiosity found sufficient evidence for the Science Team to conclude that in ancient times, the Mars environment was habitable for microbial life as we know it on Earth. As she climbed Mount Sharp, evidence led to a similar consensus that the development of Mars was dynamic, with extreme variability in the extent of surface water. As evidence accumulates, it may be clear that such dynamic behavior was Mars-wide. On Earth, extreme climate change in the past has repeatedly led to the sudden extinction of a large fraction of species as well as the accelerated evolution of new species. Is it possible that climate changes on Mars were too rapid for the sustained presence of life? Or not stable enough to allow life to emerge?

Curiosity did find two possible biosignatures. It turned out that small fragments of organic compounds could be identified in samples only a few centimeters under the surface in an area where sulfur was present to help preserve them. However, inorganic genesis was possible. But the goal of finding a good place to look was achieved.

Further, a great deal was learned about how methane is generated on Mars. On Earth, most methane is generated by living organisms or by decay processes of living organisms. Detection of methane in the Mars atmosphere has been seen from Earth and from Martian satellites, but always for short intervals and never verified. In the first year on Mars, Curiosity's sensitive instrument SAM sniffed the Mars atmosphere regularly and looked for methane, but did not sense levels of methane significantly above background level. A paper was written making that clear. Science media reported with a headline "No methane on Mars", implying that previous sightings were false. The SAM instrument has a different mode than had not been used up to that time because it takes a lot of time and power. In that mode, nitrogen and water vapor are removed from each sample of the atmosphere and the remaining gas is tested for methane. The sensitivity is thus magnified, and in this way, methane became regularly detected above the background level of the new mode. The newly detected methane turned out to depend on seasonal variation. Sporadically, much large plumes of methane were also measured. These disappeared in a day or two, although models suggested they would persist longer. Recently, a very large plume was sensed not only by Curiosity but also by a Mars satellite passing near Gale crater. Finally, independent verification legitimized the past sporadic observations. So a common product of life on Earth has been observed on Mars. As in the case of fragments of organic molecules, methane can be generated by rocks with certain minerals and water, so the methane is not indisputable evidence of current or ancient life on Mars.

In summary, a great deal has been learned about how methane is generated on Mars in sporadic plumes of the gas. They are not only patterned by seasonal changes in surface temperature but also have been detected in both in situ and orbital instruments in similar places at the same time. Although the gas disappears unexpectedly quickly, theories have been generated to explain this phenomenon (see the results section of Chap. 10).

The goal for Curiosity is not to detect life. It would be excellent if she did, but she does not have instruments that have a high probability of doing so. To detect past life, we would need to detect structures that could only have been made by living organisms or very complex hydrocarbons that could not be explained by inorganic processes. There are also arguments made that depend on biological sorting of isotopes and a property of molecules called chirality that distinguish life forms on Earth. At this time, if life has been present and pervasive on Mars, the way to have a reasonable chance of detecting it would be to return a diversity of samples to Earth for examination. That task would need to be done by other missions.

The realistic goal for Curiosity is to find Martian environments that would be likely to preserve life if it were there. Curiosity has contributed to that objective by finding pervasive evidence of processes that deposit calcium sulfate in veins in rock. Sulfur is said to be useful in preserving the complex hydrocarbons of life.

Environmental Records

[MSL Mission and Science Investigation] *An essential point that Earth also teaches us is that in the search for signs of early life a null result is a not always a disappointment. Whatever may be lost in terms of insight into possible paleobiologic markers may be gained by an equally rich reward into the processes and history of early environmental evolution. Studies of Earth's Precambrian sedimentary record have revealed secular changes in the oxidation state, acid-base chemistry, and precipitation sequence of minerals in the oceans and atmosphere (D.J. Des Marais, Isotopic evolution of the biogeochemical carbon cycle during the Precambrian. Rev. Mineral. Geochem. 43(1), 555–578 (2001). doi:10.2138/gsrmg.43.1.555; A.H. Knoll, The geological consequences of evolution. Geobiology 1(1), 3–14 (2003); R.M. Hazen, D. Papineau, W. Bleeker, R.T. Downs, J.M. Ferry, T.J. McCoy, D.A. Sverjensky, H. Yang, Mineral evolution. Am. Mineral. 93(11–12), 1693–1720 (2008). doi:10.2138/am.2008.2955). Knowledge of an equally informative environmental history may also be uncovered on Mars. The evolutionary path of surface environments on an Earth-like planet that lacked a biosphere would make a highly desirable comparison to Earth in order to understand better the unique aspects of our own planet's history. These records of environmental history are also embedded within the same kinds of rocks and minerals that may also preserve the calling cards of biology. Therefore, an MSL mission that focuses on understanding mechanisms of potential biosignature preservation will also insure that we capture the record of early Martian environmental processes and history.*

[CB Comment] The evolution of Earth's environment has been profoundly influenced by the presence of life. Think of the banks of coal deposited by palm fronds for example, or the transformation of Earth's atmosphere by algae and sunlight. If there was life on Mars, a comparison of the effects of that life on the environment with Earth's own evolution would be enlightening. If there was no life on Mars, then the comparison would be what happens to a rocky planet with and without life. Either way, the record of the early environment is written in the sedimentary rocks, and it is up to Curiosity to help us read it. In fact, in the last few years after landing, the mission has been very fruitful in doing exactly that.

[MSL Mission and Science Investigation] *This approach holds both the hope and promise of Mars Science Laboratory. The hope is that we may find some signal of a biologic process. The promise is that MSL will deliver fresh insight into the comparative environmental evolution of the early stages of Mars and Earth. That alone is a valuable prize. MSL was specifically designed for this purpose and the MSL team has a lot going for it: veterans of years of previous rover operations permeate the engineering and science teams; strategic decision making has already benefited from stunning high-resolution image* *datasets obtained by the HiRISE camera on MRO, as applied to both drive-related terrain assessment at Gale crater and refinement of scientific objectives; and the rover itself will be the most capable robot ever sent to the surface of another planet.*

[CB Summary] Sedimentary rocks tend to lay horizontally, each strata rising in the order of the time of their deposit. On Earth, fossils, which are structures of life, tend to be more complex with time as life evolves—an important characteristic of life—and are often found preserved in sedimentary rock. The orbital observations of sedimentary strata in Gale crater was one reason it was chosen as a target.

Management of the MSL Science Team

[CB Comment] The next section of the body (2.1 Mission Summary) was covered in the first chapter of this book. The following section (2.2 MSL Science Team) described the team's structure and addressed the problems of establishing discipline in what would be the largest number of instruments and tools to be operated in the history of planetary exploration. The listing of scientists involved in the enterprise, including the analysis and documentation of the streams of data, the diversity of institutions they represent and the variety of roles they are assigned, can be found in Appendix: The MSL Science Team.

The "MSL Science Team Rules of the Road," a supplement to the published paper, establishes the way in which the members of the MSL Science Team should work with each other, stating the three principles in the box below. These principles have been important in the past in resolving some of the conflicts that occur because of scarce resources and the need to make choices.

Individuals can Meet Personal Goals While Cooperating to Maximize Achievement of Team Goals

[MSL Science Team Rules of the Road]

Meeting the scientific goals of the project will require coordinated interaction among all these participants (e.g., data sharing, interactive and interdisciplinary data analysis and interpretation, joint publications). Moreover, if this coordination is well conceived from the start, it can significantly influence the success of the project by encouraging opportunities for interdisciplinary results and discoveries and by maximizing the impact of the results of the project. While encouraging these interactions, the project must also encourage individual creativity and initiative and find ways to allow all members of the project to benefit appropriately from the scientific successes of the MSL.

[CB Summary] The MSL Project Science Group (PSG) The need for an orderly process to make difficult decisions is greater with Curiosity than a less complex mission. This need is addressed in several ways. One step is to separate decisions between those that are strategic and those that are tactical. Some of the strategic decisions had been made in the choice to go to Gale crater and not one of the other three potential targets. Gale crater was targeted because it was deep but also had a central peak that was even higher than its rim. The orbital data indicated a variety of sedimentary rocks, which are good for recording changes in water flow and for preserving organic molecules. The strategic decisions are made by the MSL Project Science Group (PSG).

[MSL Mission and Scientific Investigation] The list of MSL PSG members is shown in Table 11.1.

Table 11.1 Mars science laboratory project science group

Name	Role	Affiliation
John Grotzinger	MSL project scientist	California Institute of Technology
Michael Meyer	Mars program scientist	NASA Headquarters
David Blake	PI, CheMin	Ames Research Center
Kenneth Edgett	PI, MAHLI	Malin Space Science Systems
Ralf Gellert	PI, APXS	University of Guelph, Canada
Javier Gómez-Elvira	PI, REMS	Centro de Astrobiología/INTA, Spain
Donald Hassler	PI, RAD	Southwest Research Institute
Paul Mahaffy	PI, SAM	Goddard Space Flight Center
Michael Malin	PI, MARDI and Mastcam	Malin Space Science Systems
Igor Mitrofanov	PI, DAN	Space Research Institute, Russia
Roger Wiens	PI, ChemCam	Los Alamos National Laboratory

[CB Summary] The PSG is co-chaired by the MSL Project Scientist and the Mars Program Scientist. The members are the 9 PIs of the 10 instruments (Michael Malin is PI of two instruments, MARDI and MastCam). The Charter of the PSG is defined in the body document by the paragraph in the following box:

Charter of the PSG
[MSL Mission and Scientific Investigation]

The primary function of the PSG is to advise the Project on optimization of mission science return and on resolution of issues involving science activities.

During landed operations, the PSG will have an important role providing strategic guidance to the Science Operations Working Group (that subset of MSL science team members on shift making tactical decisions on any given sol of the mission).

All MSL Science team members are expected to adhere to the Rules of the Road and any future updates approved by the PSG.

[CB Summary] Other Members of the MSL Science Team To assist the PI, other members of the science team associated with the instrument are named by the PI, who may designate some of them Co-Investigator. There are also Science Collaborators named by the PI. JPL engineers who have been assigned to the interface of an instrument with Curiosity are designated as Science Investigators. There are other members of the Science Team that are called Associated Scientists. There are also Collaborator Scientists who are associated with members of the Science Team. The individuals assigned these diverse roles are listed with their roles in Appendix: The MSL Science Team.

The cooperative nature of the MSL Science Team, guided by the preceding section, was maintained throughout the exploration. It was not always easy because of the complex problems of interpretation that were presented by the nature of Mount Sharp and its complex history. At times, separate investigation teams were formed to develop competing explanations for unexpected observations. Also, downtime of specific instruments or engineering equipment complicated competition for resources like power, time in sunlight, and safe path availability. Fortunately, the long duration of the mission permitted discussion among many scientists over the publication cycle and even long campaigns to gather unplanned information. Sustained goodwill allowed resolution of many issues by compromise and acceptance of creative new ideas.

Gale Crater Field Site

The selection of Gale crater as a target site was very carefully formalized. It was a major factor in the success of the mission (Table 11.2).

[MSL Mission and Science Investigation] *The PSG co-chairs (Project Scientist and Program Scientist), in consultation with the PIs determined that analysis of the landing sites would be aided by the involvement of the MSL Science Team, who would be intimately familiar with the instruments and objectives of the mission. These discussions became instrumental in defining the mission-relevant landing science criteria used by the broader Mars community in framing discussion of the candidate landing sites* (Table 11.2).

It was decided to charter three PSG Working Groups, each operating under the auspices of the PSG co-chairs. The first of these was chartered to specifically look at the preservation potential of organic compounds and other biosignatures on Mars, as a function of different habitable environments thought to be present at the different landing sites as presented by the community. The second Working Group was chartered to lead comprehensive discussions and analysis of the final four landing sites. The third Working Group was charged with evaluating the traversability of the rover through sloping outcrops of the lower part of Mount Sharp. Where needed the Working Groups were inclusive of experts external to the Project and Science Team, so as to maximally enhance the return and objectivity of these efforts.

[CB Summary] The following is a description of the process of examining candidate landing sites and a listing of the attractive aspects of the Gale crater that resulted in its selection. In making decisions, members of the MSL science team referred more than once to setting a high priority to opportunities for investigation that were used in selection of Gale crater as a landing site.

Gale crater scored high on Diversity because of the distinct layers of different minerals; clays, sulfates, and cemented fractures as seen from orbit. In Context, it bordered two distinct regions, a highland area to the south and the dichotic Borealis Basin to the north. As for Habitability it had clear signs of water flowing over the edge of its crater and into the depth of the crater from both north and south. Because the floor of Gale crater is so far below the surrounding surfaces, the upward pressure of groundwater was strong below the crater floor. The possibility of lakes forming and then evaporating suggests a high preservation potential.

A mobility study promised a traversable path up the central mountain. Some paths could have turned out to be blocked, but alternate paths existed to allow reaching the key mineral layers.

Table 11.2 Four major mission relevant landing site science criteria **[MSL Mission and Science Investigation]**

Criteria	Description
Diversity	A site with a variety of possible science objectives will ensure a greater chance for scientific success. Examples: multiple and differentiated science targets, multiple types of evidence (e.g., morphologic and geologic), variety in mineralogy or styles of stratigraphic expression
Context	A site that can be placed in a larger, more regional context will ensure a greater depth of scientific understanding. The regional context provides constraints on past processes that led to the environments being examined locally. Locally derived results can, in turn, be extrapolated regionally or globally
Habitability	Sites with orbiter-derived evidence for habitable environments can be assessed to make specific predictions that will guide the exploration strategy for MSL. Particular high-priority geologic targets can be identified that can be accessed, interrogated, and interpreted by MSL
Preservation	Sites with a higher potential for preserving evidence for past habitable environments will ensure a greater chance of scientific success. Using terrestrial analogs, sites can be assessed for the particular physical and chemical conditions that retain mineralogic, chemical, or morphologic evidence

[CB Comment] Visit to the Greenheugh Pediment In March 2020, crossing the Greenheugh Pediment to reach the sulfate-bearing unit more quickly than following the planned path was considered. Despite the need for very high angles of tilt, the rover drivers have successfully brought Curiosity up on top of the lip of the pediment. The views are spectacular. Will they be able to proceed? As it turned out, after examination of a mosaic taken by the MastCam, Curiosity turned around and left the pediment. A mobility study was not encouraging, so Curiosity returned to the originally planned path. However, the visit to the pediment was scientifically productive with both remote and contact instruments, including a sample of a sandstone rock.

The next two sections of the published body document cover material on the rover and instruments that is addressed in earlier chapters of this book. Two figures not covered in early chapters of this book are presented here to represent the contents of these sections. Figure 11.1 shows the engineering components of the rover.

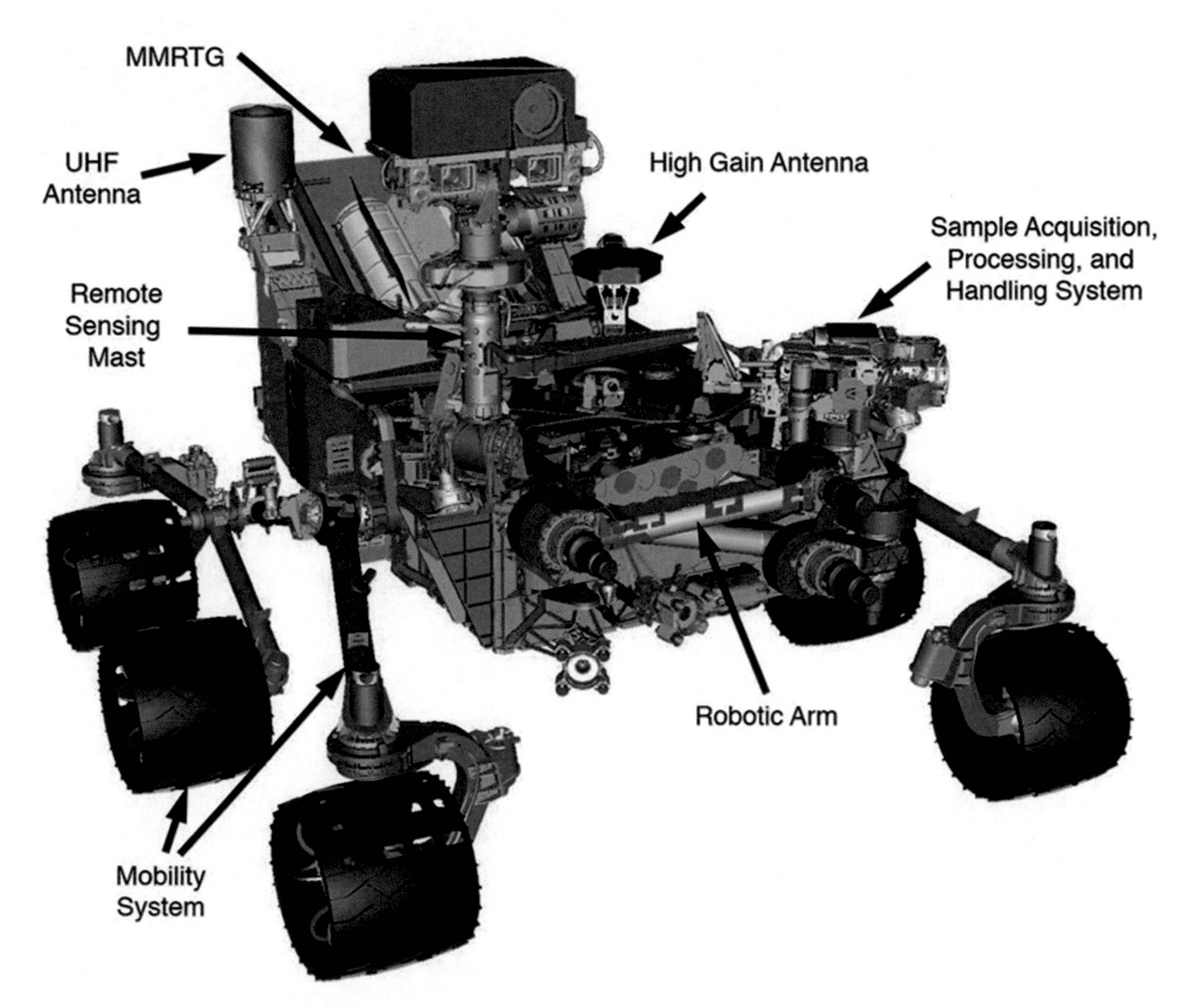

Fig. 11.1 **[MSL Mission and Science Investigation]** Drawing of Curiosity indicating some of the major engineering components. The Robotic arm is stowed (folded up) in front of the rover and the Remote Sensing Mast is not extended. Open source, John P. Grotzinger et al., "Mars Science Laboratory Mission and Science Investigation", Space Science Reviews, July 25, 2012

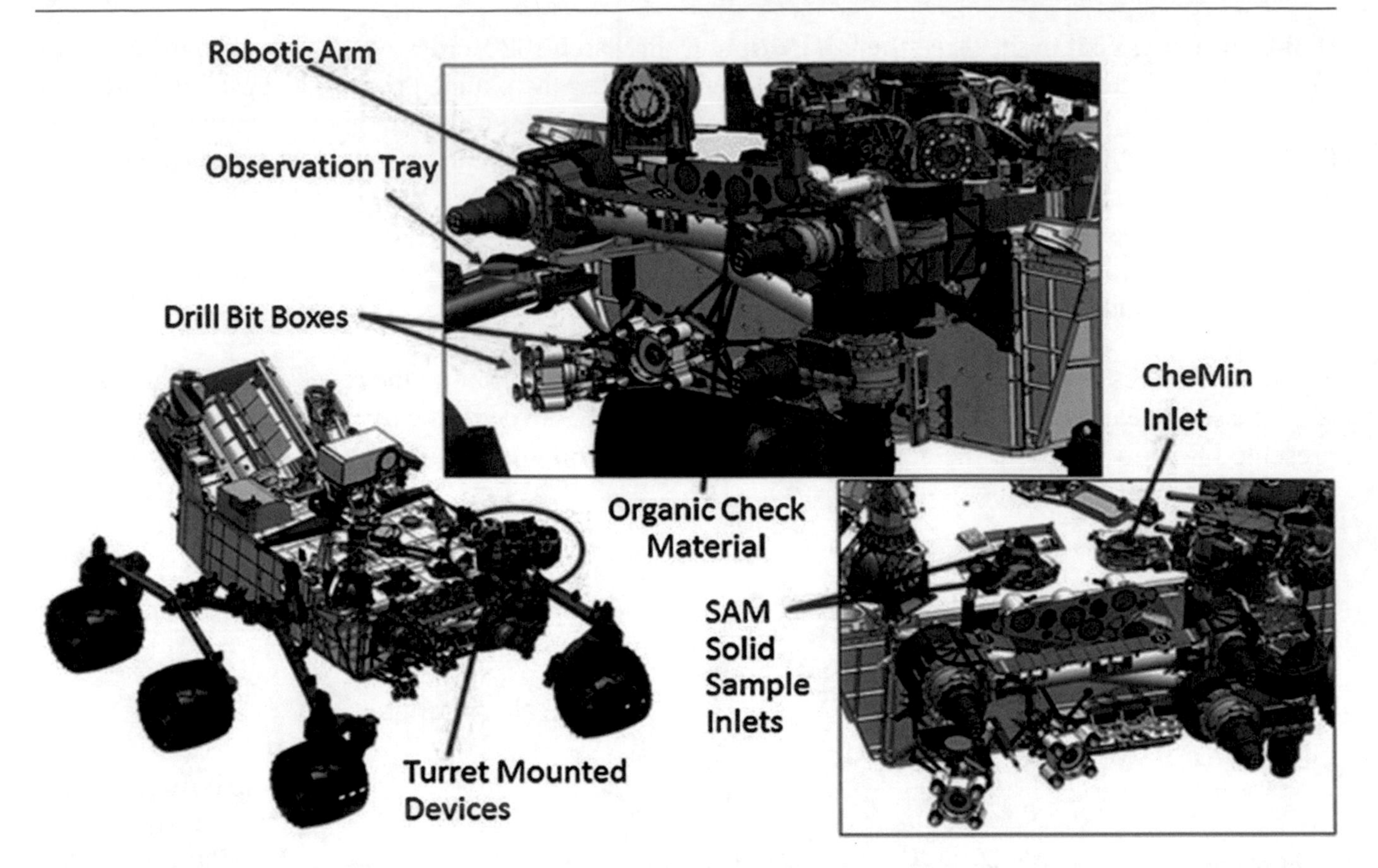

Fig. 11.2 [**MSL Mission and Science Investigation**] Diagram showing the location of Sample Acquisition, Processing, and Holding (SA/SPaH) components on the rover, including bit boxes, organic check material, observation tray, and sample inlets. Open source, John P. Grotzinger et al., "Mars Science Laboratory Mission and Science Investigation", Space Science Reviews, July 25, 2012

[CB Comment] Figure 11.2 shows some of the tools of the rover that support instruments. Each of the instruments had at least one JPL engineer assigned to be a liaison contact with the Principal Investigator and the instrument team, representing the rover's mechanical, electrical, thermal, and data interfaces with the instruments. After the landing, the rover liaison engineers assisted the instrument teams.

Entry, Descent, and Landing

[MSL Mission and Science Investigation] *Entry, descent, and landing activities occur within ~15 minutes prior to touchdown on Mars. MARDI acquires its data set from moments before heat shield separation through touchdown (<2 minutes) and a few seconds thereafter. For landing, MSL uses a propulsive descent "sky crane" to lower the tethered rover beneath it onto the Martian surface, setting its wheels directly on the ground. After rover landing, the connection with the descent stage is severed and the descent stage flies away to fall elsewhere, 150 m or more away from the rover. The rover will touch down in late southern winter (L_s = 150.7), between 14:50 and 15:02 Local Mean Solar Time on Mars, depending on the launch date.*

Commissioning Phase

[MSL Mission and Science Investigation] *The goal of the commissioning phase is to allow the mission to reach nominal science operations as quickly and safely as possible after touching down. Before this is possible, the rover operations team must characterize the health and behavior of the rover and instruments once interacting with the Martian environment. The first ~10 sols will be dominated by critical hardware deployments (e.g., the mast and mobility system), installing the flight software version used for surface operations, and spacecraft and payload checkout activities. After this initial characterization phase, mast-mounted and monitoring instruments may begin performing nominal science as operational time, power, and downlink data resources allow. The next few 10s of sols will involve checkouts and first-time uses of more advanced capabilities, such as the robotic arm and other sampling hardware. These activities will be coordinated with strategic science decisions such as whether to drive out of the region contaminated by the landing engines' effluents, to fully enable contact science, to acquire a sample of soil, rock, or organic check material, or to focus on traversing toward other scientific targets.*

[CB Comment] There was an interesting discussion over the first pictures from the hazcams to be downloaded. Some wanted a rear-facing hazard camera image to show stones first (Look! we are really on the ground!) and some wanted a front-facing hazard or NavCam to show Mount Sharp on the horizon (Wow!). At the actual landing, images were taken and sent to Earth front and back within minutes of touchdown. They were blurry and fuzzy with dust thrown up on the protective lens cover on the camera, which had not yet been removed. However, one picture was taken within seconds of the landing and showed an unsettled puff of dust that had been thrown up by a descent stage's engine. The commissioning description continues below.

[MSL Mission and Science Investigation] *These activities will be coordinated with strategic science decisions such as whether to drive out of the region contaminated by the landing engines' effluents, to fully enable contact science, to acquire a sample of soil, rock, or organic check material, or to focus on traversing toward scientific targets.*

[CB Comment] Even before landing, the MARDI camera was taking pictures during powered descent and storing them for transmission after landing. These were used, as planned, for determining the exact landing location within the target area. An early strategic activity for the PSG is to use the location of the landing to determine where to set the early targets for science activities. Four choices were discussed: go northeast to the delta from Peace Valley, go southeast to low-lying Yellowknife Bay, go south directly to Mount Sharp (crossing the Bagnold Dunes), or go southwest to pass the dunes and go through the Murray Buttes to Mount Sharp. The choice was to go to Yellowknife Bay and west around the dunes toward Mount Sharp. This turned out to be an excellent choice!

Surface Operations

[MSL Mission and Science Investigation] *MSL's primary mission spans one Mars year (669 sols or 687 Earth days) after touchdown. Science team activities will terminate six months after the end of the surface mission, whether it ends after one Mars year or after any number of extensions. Nominal science operations will occur throughout this period with a few exceptions—namely, the commissioning phase right after landing, a ~30-sol period of minimal operations centered on superior solar conjunction (18*

April 2013), ~10 sols dedicated to software updates throughout the primary mission, and a few other maintenance activities.

MSL is intended to be a discovery-driven mission, with the science operations team retaining flexibility in how and when the various capabilities of the rover and payload are used to accomplish the overall scientific objectives. One major partition in the rover's activities is between driving and "sampling," where the latter represents a series of environmental, remote sensing, and contact science measurements may then lead to the acquisition, processing, and analysis of a sample of rock or soil in the analytical laboratories. The proximity of the specific touchdown location to targets of scientific interest within the landing ellipse, and to Mount Sharp itself, will influence the ratio of driving to sampling in the early mission.

Science activities on any particular sol are governed by a number of constraints that are measured or predicted for that sol, such as the Earth-Mars geometry and local time phasing, timing of telecom windows, downlink data volume capability, the time profile of energy available for science, and any thermally driven operational constraints or energy needs of the payload, rover subsystems necessary for payload operations (e.g., robotic arm actuators), or the rover. Science activities generally require more power than is available from the RTG and rely on drawing down the rover batteries. (This is also true for many engineering activities.) Battery capacity, RTG output, overnight battery recharge, and management of the state-of-charge over multiple sols, are all critical to science (and engineering) planning. The thermal limitations, including significant time and energy required to heat mast, arm, and mobility actuators, vary with both time of day and season. Most science activities will occur during daylight hours on Mars (Fig. 11.3). Throughout the mission the rover itself operates on Mars time. Commands for the sol's activities are sent via the overnight orbiter telecom pass or direct-from-Earth at local mid-morning on Mars. The rover will complete its tactical science activities (i.e., those that influence planning for the next sol) in time to return the data via an orbiter telecom pass in the midafternoon. During the early portion of the mission, the operations team will synchronize its efforts to Mars time. Between midafternoon and the next morning on Mars, the science operations team on Earth will assess the downlink, plan the next sol's activities, and prepare the commands. Data that are not essential for next-sol planning will be returned during the overnight orbiter telecom pass. This basic framework allows

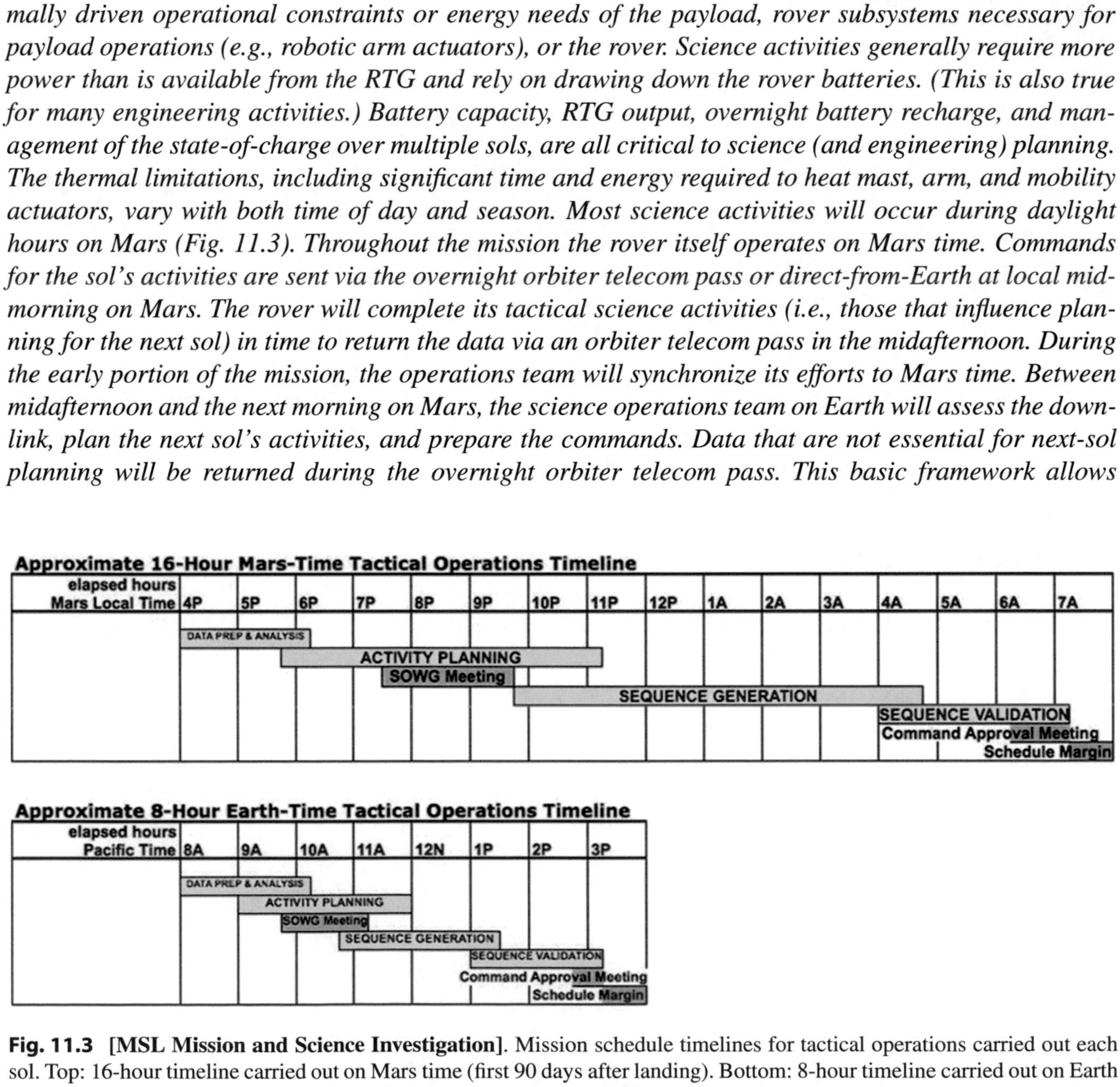

Fig. 11.3 **[MSL Mission and Science Investigation]**. Mission schedule timelines for tactical operations carried out each sol. Top: 16-hour timeline carried out on Mars time (first 90 days after landing). Bottom: 8-hour timeline carried out on Earth time. SOWG is the Science Operations Working Group. Source: Open source, John P. Grotzinger et al., "Mars Science Laboratory Mission and Science Investigation", Space Science Reviews, July 25, 2012.

approximately five hours for tactical science activities by the rover on Mars. Additional payload or rover operations can occur outside of this window if they are not critical to the next sol's planning. Following the first 90 sols the operations team will revert to an Earth-based time schedule (Fig. 11.3).

During winter, the time available each sol for science operations may be reduced because of the need to use a greater share of energy to heat the rover actuators. Also, the largest actuators may not warm sufficiently until after the afternoon orbiter telecom pass. For this reason, winter operations may use every-other-sol commanding for more than 12 out of every 36 sols.

Mission Operations After Landing

[MSL Mission and Science Investigation] *The first ~90 days of the mission is accomplished with all operations participants on site at JPL, with personnel working in shifts synchronized to Mars' 24.6-h day and on duty around the clock, seven days a week. A description of how Mars Time operations worked for the Mars Exploration Rover mission is given in D.S. Bass, R.C. Wales, V.L. Shalin, Choosing Mars time: analysis of the Mars Exploration Rover experience, in IEEE Aerospace Conference, 5–12 March (2005), pp. 4174–4185, paper #1162. doi:10.1109/AERO.2005.1559722 and A.H. Mishkin, D. Limonadi, S.L. Laubach, D.S. Bass, Working the Martian night shift—the MER surface operations process. IEEE Robot. Autom. Mag. 13(2), 46–53 (2006). doi:10.1109/MRA.2006.1638015. Operating on Mars Time and extra staffing to cover the tactical uplink process will allow the extension of the tactical timeline from the normal 8 hours to a two-shift, 12 to 16 hour timeline (Fig. 11.3). A major objective of this period is to develop the capability to complete the tactical one-sol turnaround process in 8 hours or less, by increasing efficiency.*

After a portion (or all) of this initial period, the flight team begins transitioning to operate via a distributed operations network, with the central hub at JPL. This enables the remote science teams to work from their home institutions for the long duration of the mission, interacting via internet and phone teleconferencing. The start time of the prime shift on Earth will track Mars time, sliding forward from 6 AM until it reaches 1 PM (Pacific). After this point, the downlink from Mars arrives too late in the day on Earth to allow commands to be generated before a reasonable end of shift. In these cases the ground cycle is postponed until the next available Earth shift. From a tactical standpoint, every other sol is lost during this period. However, science activities can be performed by the rover on every sol as long as they can be planned in advance and/or their results are not required immediately for future planning. This period of every-other-sol (or multiple-sol) commanding is expected to span about 12 sols of every 36-sol Earth-Mars phasing cycle. After the first six months of operations, the operations team will support 5 day per week tactical operations on Earth time, with multiple-sol rover plans prepared for weekends and holidays.

Strategic Planning

[The MSL Supratactical Planning Process] *The Strategic planning process addresses the long-term aspects of planning, including development and testing of first-time activities, planning science campaigns, and long-term management of rover resources and constraints.*

[CB Comment] As mentioned earlier, strategic planning is the responsibility of the Project Science Group (PSG). The group has a responsibility to plan the path of the rover in a way that the mission's goals are reached efficiently and to ensure that the highest priority goals can be met. Consultation is necessary with mobility experts to assure that the chosen paths are trafficable or that alternate portions of the path are available. As the mission progresses and knowledge of the environment increases, the PSG revises the path to maximize scientific results.

The strategic plan for the conditioning phase after a successful landing had been pre-planned. When conditioning came to the mobility system, a strategic choice was needed. Now that the specific location was known within the target area, should Curiosity go on a long traverse westward to reach the planned ascent path to climb Mount Sharp, with its sequence of new mineral goals, or go southeast on a shorter path to Glenelg where there was a conjunction of three geologic units? They went toward Glenelg – a fortunate choice, because it led them to a streambed at Yellowknife Bay, where they achieved Goal I, sufficient evidence for ancient habitability there.

Starting along the path to Glenelg for mobility conditioning and leaving the area contaminated with exhaust from the rockets of the descent stage, they stopped at Rocknest to continue conditioning of the science and instruments. Curiosity took the first scoop sample there of Bradbury soil and passed it to SAM for analysis. By the end of the initial 90 days of operations as well as conditioning, the strategic and tactical processes had started working as planned, and significant science had been performed. The rover had covered 490 meters, the cameras had produced excellent images, and Curiosity had taken its first selfie. The new supratactacle process was evolving.

In Yellowknife Bay, the first major area to be explored, interesting features were seen in the cameras and visited. As a result, priorities changed as they came close to smaller but interesting features. The result was a confused path, with some features visited multiple times. As the mission progressed, it became a custom to do a walk-about when entering a new area; that is, to systematically cover the length and breadth of the area before selecting targets. This process was found to increase productive efficiency.

Tactical Planning

[MSL Mission and Science Investigation] *The Tactical planning process is the truly reactive aspect of planning, responding to data received from the rover each sol. The timescale for Tactical Planning is typically one sol.*

[CB Summary] Tactical decisions are the responsibility of the Science Operations Working Group (SOWG) each sol to plan the science activities for the next sol (or in some cases two or three sols). The SOWG chair is chosen by the members and is approved by the PSG. The chair of the SOWG reports issues that cannot be resolved by the group to the PSG.

[MSL Science Corner] *Members of the SOWG represent the following themes*

Organic Geochemistry and Biosignatures	*Chemical and isotopic composition of organic compounds in solid and gas samples and other elements/compounds of relevance to habitability* *Textural, chemical, mineralogical, and isotopic biosignatures*
Inorganic Geochemistry and Mineralogy	*Chemical, mineralogical, and isotopic composition of rocks and soils*
Geology	*Bedrock geology, geomorphology, and stratigraphy* *Rock and soil textures* *Rock and soil physical properties*
Atmosphere and Environment	*Meteorology and climate* *Distribution and dynamics of water and dust* *Solar, UV, and high-energy radiation* *Atmospheric chemical and isotopic composition*

There is a thermal representative to monitor that the instruments are in their temperature ranges, an uplink manager, and a downlink manager that assure that the data requirements are within the allowance.

[CB Comment] Early during the Rocknest and Yellowknife Bay Explorations, the SOWG realized that sampling by scooping or drilling and waiting for the results of SAM analysis was taking longer than expected. These activities used a great deal of power and time, especially if the results of SAM analysis were reviewed before starting the next sample. As the mission progressed and operations become more complex, especially as unexpected environmental hazards damaged the wheels, it became difficult for an SOWG meeting to plan for operations that would take two or three sols to carry out, depending on the type of activity.

[CB Change] Since the Supratactical Planning Process has been implemented, the SOWG has been guided by a skeleton plan for multi-sol activities from the newly created Supratactical Planning Process.

The Tactical Planning Process

[CB Summary] Daily SOWG Meeting Schedule Initially, for the first 90 sols after landing, when all active planning personnel were physically at JPL, the original Tactical Process meeting took 16 hours. This was also the time between the download of data from the end of a sol to the start of Curiosity's day for uploading the commands for the next sol. The time available to get yestersol's report and plan the operations for tomorrowsol was equal to the time needed to create the plan. This required the work time to slide with Mars time, starting 39 minutes later each Earth day. Obviously, locking the work time to Mars time caused great stress on the members, considering both health and family, especially on weekends.

[CB Change] Part of the problem was that harmonizing the long-range Strategic Planning with the short-range Tactical Planning was taking time from the Tactical Process. Inserting the Supratactical Process with additional personnel removed the burden of harmonization from the Tactical Process, ultimately allowing them to complete their work in 8 hours. The PSG changed the schedule, allowing them to meet at the same time for a number of sols and then move to another stable time. It also allowed for planning, on occasion, of 2 or sometimes 3 sols, additionally relieving the problem.

[CB Comment] Of course, with so many Solar System scientists in the MSL Science Team, a few will want to see an eclipse of the sun, even if it is only partial (see Fig. 11.4).

Sol Types One of the actions of the Tactical Process is to designate, according to the skeleton plan, the type of the next sol. The sol type varies in the instruments and rover tools that are used to achieve the results of the next sol. The chart in Fig. 11.5 (from the original published document) describes the sol types. The sol types for a series of sols may be specified in the skeletal plan provided by the Supratactical Working Group to the Science Operations Working Group.

The purpose of defining sol types is to provide focus to the activities that can be carried out efficiently in a given sol. For example, Traverse to a new location will require a drive. The longer the drive, the less power will be available for other functions. There are two combination sol types, Dilution/Sampling and Analysis/Observation tray. The dilution, sampling, and analysis functions use a lot of time and power. The observation tray uses a lot of time.

Fig. 11.4 A subset of the Curiosity tactical planning team (SOWG) took a short break from planning during sol 1793 to view the solar eclipse of August 21, 2017 from the roof of building 264 at JPL. Notice that there are fewer people than Curiosity has instruments and tools. Additional members of the SOWG were in contact electronically from other locations. (Image courtesy of NASA/JPL-Caltech)

	DAN	Mastcam	ChemCam	MAHLI	APXS	CheMin	SAM
Traverse	survey during drive	corridor and end-of-drive panoramas	pre- and post-drive survey				
Recon		panoramic imaging	survey	single target	long integration		
Approach	high-precision measurement	workspace imaging	pre-drive survey	single target	short integration		
Contact		pre- and post-brush imaging; multispectral	analysis of contact target	pre- and post-brush imaging of contact target	pre- and post-brush analysis of contact target		
Dilution / Sampling		support imaging				sample analysis (begin)	
Analysis / Obs Tray		support imaging		sample analysis	sample analysis	sample analysis (end)	sample analysis

Fig. 11.5 **[MSL Mission and Science Investigation]** This is a chart of instrument activities on MSL soil types. The left column of this chart shows the types of sol. The remaining columns are the scientific activity that may be used for that type of sol. Open source, John P. Grotzinger et al., "Mars Science Laboratory Mission and Science Investigation", Space Science Reviews, July 25, 2012

Preparation activities are needed at the beginning of each of these activities, and specific observations and data downlink are needed at the end. Dilution is a process of removing contamination. Curiosity processed and analyzed five scoop samples to flush out residual organics. Then the first two drill samples were taken close together to check for different levels of contamination. For Contact sol types that use the arm and turret to position instruments, only one instrument can be used at a time.

With the exception of the sampling, analysis, traverse for long distance, and imaging for large mosaics, activities can often be carried out within one sol. But for these important activities, a new human group with a longer view than one to three sols is needed to orchestrate the tactical decisions.

Supratactical Planning Process

[The MSL Supratactical Planning Process] *The Supratactical planning process bridges the gap between Strategic planning and Tactical planning, and incorporates certain aspects of predictive as well as reactive planning.*

[CB Summary] The multi-sol planning problem of the SOWG was seen as stemming from the two-step decision-making method—strategic and tactical—that did not allow for multiple sol sequences to be planned without waiting for the results from the first sol in the sequence. The need for this was the increased complexity of Curiosity's science instruments, particularly in regard to the time and power requirements for the sampling process and the SAM suite of instruments.

The solution was to change the planning process to insert an intermediate planning step called the *Supratactical Planning Process* between the strategic and tactical levels, with a new team of members selected from personnel that had been involved in development of the rover's power system and instrument interfaces. An example of this team's role was holding back power requirements for one sol to charge the battery so there would be more power available for the next sol. This process allows for sustained activity for many sols and peak power (MMRTG plus battery) for intermittent sols.

This modified the original decision process that had been based on experience with the Mars Exploration Rover mission. The original two-step process was not sufficient for planning of the more complex Curiosity rover, with 10 science instruments competing for power, time in sunshine, arm and turret, etc. A paper on the new MSL Supratactical Planning Process was delivered by JPL's Debarati Chattopadhyay et al. at the 13th International Conference on Space Operations. The three-step planning process is shown in Fig. 11.6.

[CB Comment] The paper "MSL Superatactacal Planning Process" mentioned above concluded:

> **[The MSL Supratactical Planning Process]**
> *Given the lessons learned from MSL, it is likely that future missions which have complex payloads, and interactions with a planetary surface or other rapidly changing environment resulting in a time-constrained tactical timeline could benefit from a Supratactical process. It is worth considering early in the mission development whether this type of process could add value to the mission.*

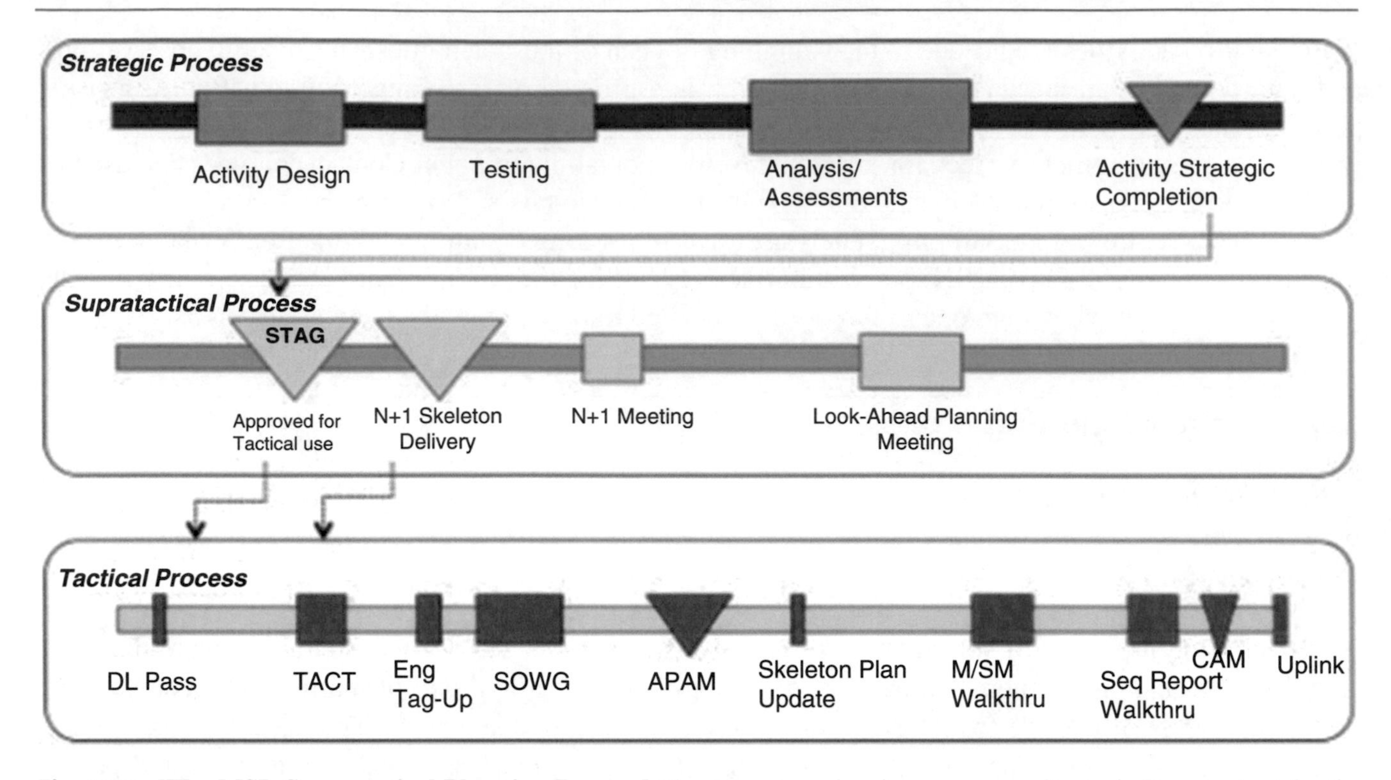

Fig. 11.6 **[The MSL Supratactical Planning Process]** The three-step planning processes: Strategic Process (top) by the MSL Project Group (PSG), Supratactical Process (middle) and the Tactical Process (bottom) by the Science Operations Working Group. **Key**: STAG is the SupraTactical Approval Gate, N+1 is the next sol in a potentially multi-sol skeletal plan, DL Pass is DownLoad Pass from the orbital data relay (results of the previous sol), TACT is the Tactical ACTivity Planning Meeting, Eng. Tag-Up is meeting of SOWG Chair with Engineering Leads, SOWG is the Science Operations Working Group meeting, APAM is the Action Plan Approval Meeting to approve the rover and science action plan, M/SM is a meeting of Mission and Science members to walk through the planned sequence for the next sol and relate them to the mission goals, Seq. is Sequence of Payload Uplink events, CAM is the Command Approval Meeting. Open source, AIAA Conference on Space Operations 2014, Chattopadhyay, et al., The MSL Supratactical Process, DOI: 10.2514/6 2014-1940

Estimated Mission Performance at Gale Crater

[CB Summary] How do you measure performance of a planetary mission? Before landing, the MSL Project Group decided to select as measures of performance the distance traveled by Curiosity and the number of samples taken and analyzed in the first Mars year after landing. Of course, it was really knowledge gained that mattered. Both the questions answered plus the new questions raised will outlive the mission itself. Nevertheless, performance figures could be estimated before launch as measures of design success or failure. 18-km distance traversed and 11 samples taken in one Mars year (668 sols) were deemed best-case measures that assumed that there were no interruptions due to environmental problems or equipment failures resulting in down time.

The MSL survived launch, cruise to Mars, the entry, descent, and landing in the planned target area, and on-schedule communication with the orbital escort spacecraft. After one Mars year (about two Earth years) in the Martian environment, it was still traveling and still had full capability for sampling and analysis. Curiosity had traveled 8 km and had taken 5 scoop samples and 3 drill samples by sol 668.

In terms of the selected performance numbers, things did not look too good, except that we had a healthy Curiosity. Now, after four Mars years (and still going), we have traveled more than 20 km and taken over 30 samples. So, Curiosity is getting the job done, even if it is taking longer than originally estimated. The increase in later productivity is probably linked to the introduction of the Supratactical Planning Process near the end of the first Mars year. Another factor may have been the systematic use of a walk-about survey when entering a new area.

[CB Change] In terms of answering questions, Curiosity is a tremendous success. In only half of a Mars year, she has unquestionably relayed images and sampled minerals that resolved the central question of its mission: has Mars ever been habitable? The answer is an unqualified yes. Mars once had flowing water with a nearly neutral Ph level. The mudstone rocks had the elements needed for life as we know it and the molecular composition of the rocks showed that energy sources were present. Earth life could survive now, if water was supplied and shelter provided for protection from the Mars atmosphere and the solar wind. Perchloride ions in the soil and atmosphere are toxic to some organisms and food for others. Other answers have come as well, described in the Results section of each chapter. Other questions have been raised, some of which have also been answered by the Mars Science Laboratory.

Scientific Papers Published by the MSL Science Team Members Another measure of productivity is the number of peer-reviewed published papers in professional journals. The MSL project has kept such records and the number per year (after the year of landing) and the cumulative number are shown in Fig. 11.7.

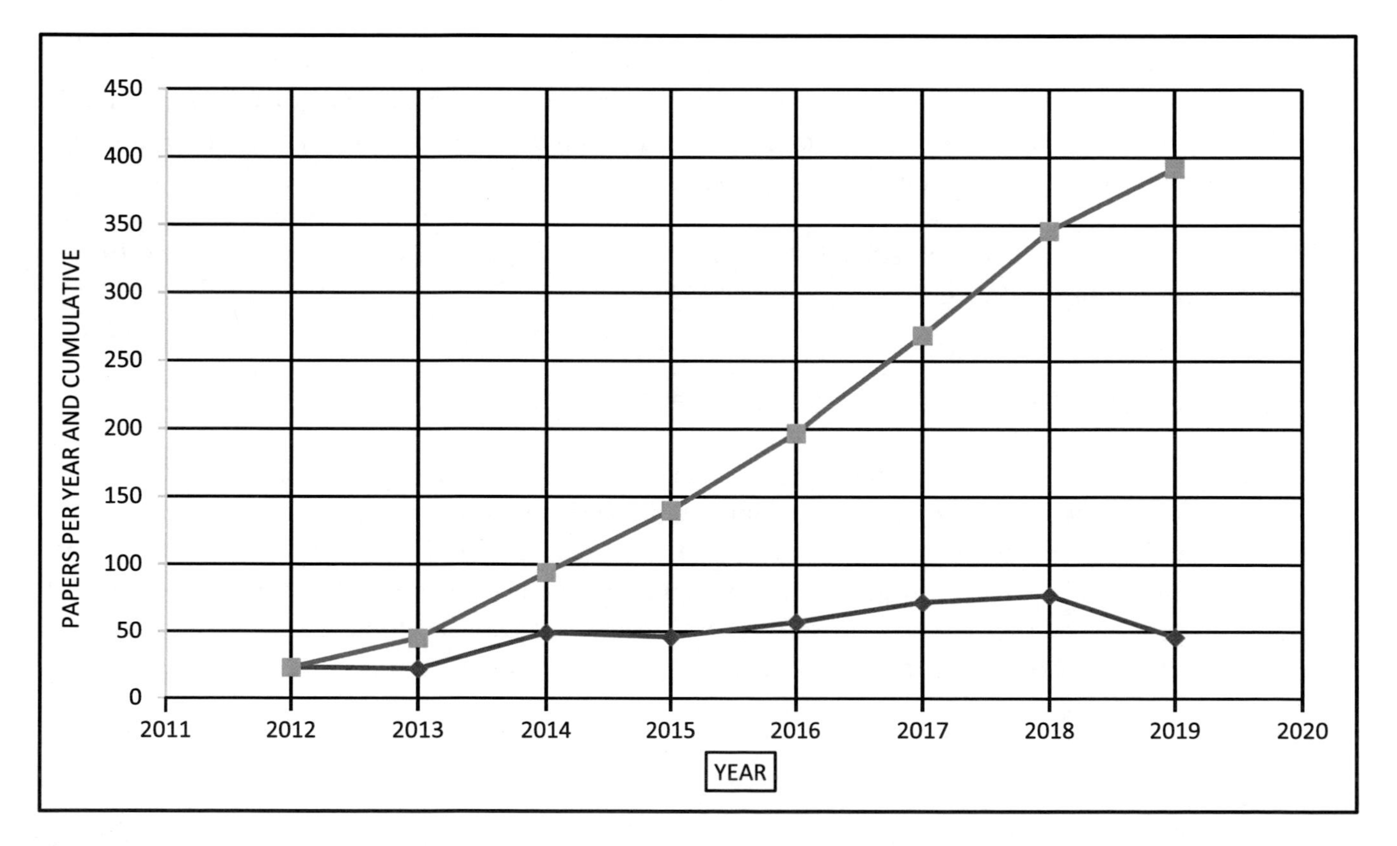

Fig. 11.7 Peer-reviewed published papers (black) from the MSL Science Team by year of publication and the cumulative number of papers (blue). Data source: NASA web page: https://mars.nasa.gov/internal_resources/840/. Graph by Charles J. Byrne

Planetary Protection

[CB Summary] The following is a summary of the extensive steps taken to comply with international standards to avoid contamination of Mars by the Mars Science Laboratory.

In the study of whether Mars has had environments conducive to life, precautions must be taken to avoid introduction of microbes from Earth by robotic spacecraft. The MSL complies with an international treaty and with NASA regulations. The following statement is a more detailed description of the methods taken to avoid contamination by Curiosity and supporting MSL units that reached the Martian surface:

[MSL Mission and Science Investigation] *NASA's primary strategy for preventing contamination of Mars with Earth organisms is to be sure that all hardware going to the planet is biologically clean. The Mars Science Laboratory mission is allowed to carry up to 500,000 bacterial spores on the entire flight system, specifically including the equipment discarded during entry, descent and landing.*

The standard of cleanliness is even stricter for portions of the rover's sample-acquisition hardware that will contact the Martian subsurface or the interior of rocks. While these components are baked to sterilize their surfaces using the same techniques as other hardware, special care is taken to prevent possible recontamination that can occur even in a cleanroom.

The Mars Science Laboratory is also complying with a requirement to avoid going to any site on Mars known to have water or water-ice within one meter of the surface. This is a precaution against any landing-day accident that could introduce hardware not fully sterilized by dry heat into an environment where heat from the mission's radioisotope thermoelectric generator and a Martian water source could provide conditions favorable for microbes from Earth to grow on Mars.

[CB Comment] Since 2013, after Curiosity's investigation of Yellowknife Bay was completed and its results understood, it became clear that extremophile bacteria could have a chance of survival under the surface, protected from the solar wind, where liquid water might be present. In light of these findings, the need for planetary protection persists for Mars.

There have been problems experienced with the sterility of Curiosity's drill bits and wheels. In both cases, remedial action was taken.

Summary of the MSL Mission and Science Investigation

The Mars Science Laboratory Mission represents an ambitious step forward in the exploration of the planet. The interplay of sequential and overlapping orbiter and rover missions has dramatically improved our understanding of the history of Mars surface environments, including those that may have been habitable by microorganisms, had life evolved on Mars. The data from Curiosity's ten scientific instruments has been entered, after calibration, into NASA's Planetary Data System.

The rest of this chapter summarizes and discusses the MSL Science Team Rules of the Road.

The MSL Science Team Rules of the Road

[CB Summary] The management structure of the Mars Science Laboratory Science Team was established by a document entitled "Mars Science Laboratory Science Team Rules of the Road," a supplement to the published "MSL Mission and Science Investigation," the subject of this chapter so far. The members of the MSL Science Team were listed in the MSL Rules of the Road and are also listed in Appendix: the MSL Science Team of this book. It is an impressive roster of about 400 experts in planetary science and instrumentation from a diverse set of institutions, both American and International. Like the body document, the Rules of the Road supplement was published on July 25, 2012, just a few weeks before the landing of Curiosity.

The document was published as a supplement to "Mars Science Laboratory Mission and Science Investigation," an article in *Space Science Review*, July 25, 2012. It is an historical document, recording what the leaders of the MSL were thinking about management of surface operations just before the landing. Topics included were making operations decisions to optimize science results, data sharing among members of the science team, and publication ethics. The document was prepared by John Grotzinger, MSL Project Scientist, Ashwin Vasavada, Deputy Project Scientist, Joy Crisp, Deputy Project Scientist, and Michael Meyer, MSL Program Scientist. Nine Principal Investigators, were listed as "concurred by." Professor Grotzinger has kindly given this author permission to include the supplement containing the MSL Science Team Rules of the Road in this book. Parts of the rules are modified in order to reflect changes that took place during the mission, specifically the creation of the Supratactical Planning Process to manage the complexity of surface operations of the many science instruments of the rover.

The Rules of the Road describes a management team called the Program Science Group (PSG): members are John Grotzinger, the Chief Scientist, Michael Meyer, the Program Manager, the Principal Investigators, with others as needed. In January, 2015, Aswin Vasavada succeeded John Grotzinger.

Note that the two management documents were published about two weeks before Curiosity landed in Gale Crater on August 6, 2012, so the personnel listed in Appendix A were organized before touchdown (they were selected in the previous November about the time of launch), although the way they would work together was still being finalized. In fact, the mode of operations changed as the mission went on and experience with the most complicated mission to date was gained.

The Rules of the Road addressed two topics. One was how the MSL Science Team worked together and with the PSG during mission operations. The other, as results were achieved, concerned data privileges and publication ethics.

Operations decisions were divided between strategic and tactical issues. Strategic plans, like broad areas to explore and planned routes, are the responsibility of the PSG. Tactical plans, like how to maximize science results in an interesting area, are the responsibility of the Science Operations Working Group (SOWG). Each such group has a theme, voluntary members, and a chair. The PSG provides oversight to the working groups and assures that their decisions support the strategic plans.

As described above in this chapter, the Supratactical Process, with additional members, was added to deal with multi-sol coordination and allow the Tactical Process to concentrate on the next sol, in consideration of the results of the previous sol.

The data privilege polices and publication policies were substantially unchanged in the course of the mission so far.

Data Privileges Policies

[CB Summary] Data sharing managed by the MSL project, consistent with the Mars Exploration Program Data Management Plan, is of four types:

- Data sharing within the MSL project
- Data release to the general public
- Release of data and discussion of interpretations through the media (print/radio/TV/film)
- Data sharing with the science community

[MSL Science Team Rules of the Road] ***Within the MSL****, each investigation should produce processed data products at their home institution or at JPL and provide them fully to the Ground Data System (GDS) to be available to the entire science team and to the engineering operations teams. In principle, this also applies to engineering data being available to the science team. This is important in guiding the Strategic, Supratactical, and Tactical planning processes so that each sol has the benefit of the results of the previous sol's activities in order to optimize the scientific return of each sol. It is the responsibility of Principle Investigators or Participating Scientists to distribute data and data products in a timely fashion to their collaborators. It is expected that data will evolve from raw, to provisionally analyzed, to validated and to archivable states. All data products will be made available to any MSL team member or collaborator.*

> *As a general rule, any MSL data products (including calibration data) will be made available to any MSL Team Memberor Collaborator.*

Data Release to the General Public *In order to engage the public, the MSL Science Team or NASA (through the MSL project office) may release interesting data or data products from each of the science instruments in a timely fashion. Unless such information has not been previously released publicly by NASA or MSL, it may only be released with the approval of the PSG or through a PSG defined process. The PIs will have the primary responsibility for representing and coordinating their teams regarding such releases. These approvals also apply to websites maintained by team members, collaborators and their institutions, as well as blogs to other web sites.*

All images downloaded from the two Navcams and the four Hazcams will be posted as rapidly as possible on a world Wide Web site hosted by JPL. Release of these images to the web will not be delayed intentionally and will not require review or approval by anyone.

[CB Comment] Image Data Transparency This policy has not only been honored but applied to the science cameras as well. It has helped project an image of transparency for NASA. The raw images have been archived and are easily available by sol or date on a webpage that can be found by a search for Mars MSL Raw.

[MSL Science Team Rules of the Road]Release of Data and Discussion of Interpretation Through the Media (Print/Radio/TB/Film) Interviews of members of the MSL Science Team should be coordinated with the JPL Media Relations Office and approved by the PSG. Requests for approval are expected to

be coordinated by the PIs, leaving several days for the approval. The PSG will work with JPL Media Relations to develop guidelines for different levels of public exposure, different levels of media training of the members, information about mission status, and special situations.

Each PI and each MSL Science Team memberi may release data from their home institution's media relations office, provided the PSG approval and JPL Mediations coordination has been completed. An important issue in releasing information is to share credit appropriately within and across PI-led teams and the PSG will take particular care in considering this issue.

Data Sharing With the Science Community By NASA policy, investigators do not have exclusive use of the data taken during their investigation for any proprietary period. However, it is recognized in the Announcement Opportunity for the MSL project that some time is required, not to exceed six months, for data products to be generated and validated. Therefore, PIs are responsible for delivery to the Planetary Data System (PDS) of Level 0 and Level 1 data (to PDS standards) no more than six months after receipt on Earth.

The documentation delivered to the PDS shall describe the higher-level products must include a complete description of the methods of generating them. A reasonably skilled end user should be able to understand the methods and reproduce the scientific results derived from the data products.

Before delivery to the PDS, no data products shall be released to the science community other than results contained in scientific publications or associated supplementary data unless the products were released to the general public as described in this Rules of the Road document or the MSL Archive Generation, Validation and Transfer Plan. This requirement must be accepted by all members of the MSL Science Team in order to protect confidentiality and openness within the team.

An exception to this rule is that relevant results may be released to selected members of the community if an unexpected situation arises and there are no existing team members or collaborators with adequate expertise are available. All such releases must be approved by the MSL project scientist after consultation with the relevant PIs or the entire PSG.

Publications

[CB Summary] It was anticipated that there would be a steady stream of results from the array of instruments and cameras that would generate scientific papers, abstracts, and talks. The responsibility for coordinating these releases lies with the PSG. The Rules of the Road discusses the role of the PSG, authorship guidelines, anticipated publications, presentations at scientific conferences, informal talks, and follow-on science. Given the large number of scientific participants, the integrated nature of most of the anticipated results (i.e., most publications will involve team members and collaborators associated with multiple PI-led investigations), and the importance that most scientists attach to obtaining recognition for their work, it is anticipated that the twin goals of effective communication and achieving equity may require delicate balance and coordination of the team and collaborators. The following is a short summary of the very detailed content of the Rules of the Road on this important topic. Reference should be made to a current version of the Rules of the Road or the Science Corner on the MSL mission webpage.

The Role of the PSG The coordination and implementation of the publication policy will be the responsibility of the PSG, including

- What papers will be written
- Authorship
- Which results will be put in the papers
- Coordination of the results of the project to the scientific community
- Balancing issues of equity and quality among the many participants in the project
- Allocation of credit for obtaining and processing data
- Allocation of credit for creativity and development of interpretations and hypotheses
- Allocation of credit for scientific leadership within the project
- Respect divergent interpretations
- Encourage the publication of minority viewpoints and multiple interpretations of the same observations
- Meet regularly to monitor the progress of manuscripts in preparation and discuss plans for future publications.

PSG Decision Process It is expected that most decisions will be made by consensus. If a decision is not supported by a consensus, the project scientist shall attempt to craft a compromise that the PSG accepts. If that is not successful, the co-chairs shall make a decision. If one of the co-chairs has a personal stake in the outcome of the decision, the other co-chair shall make it.

Authorship Guidelines Complex rules for authorship prioritize team member, collaborators, and other members of the scientific community in that order, providing they make a substantive contribution to the writing and/or the research reported in the paper.

[CB Comment] In general, the PSG has performed well, as the list of nearly 400 publications with many co-authors each attests. In reviewing the long list of responsibilities above, they were presented with certain challenges. The largest and longest lasting issue was how to handle the controversy concerning alluvial or aeolian transport and deposit of sediment. After exploration of the Pahrump Hills, opinion was turning toward lake deposits. After Marias Pass, aeolian transport gathered supporters for certain units that were newly observed there. In the summer of 2017, each of the proposals was supported by alternate peer-reviewed papers of many authors, approved by the PSG. A compromise proposal was published by 2018 with a revolutionary proposal of the Siccar Point Group that postulated the alluvial Murray Formation was deposited on the shores of Gale Lake as its surface rose in a series of wet periods. Then the younger aeolian Stimson Formation was deposited as the lake level alternately fell and rose in cycles, ultimately falling, as the climate dried and the sediment lithified. This compromise paper was authored by both groups.

The mission's scope was redirected to the extensive "Dunes Campaign' to explore aeolian processes as they continue to the present.

The Rules of the Road asserts authority of the PSG for six months after the end of MSL's exploration of Gale crater. The end may come with some random, fatal event, and also could come not with a bang but with a whimper, such as a gradual decline of power. In that case, high-power instruments will be turned off first and gradually, PIs will withdraw. In that event, it is likely that the Rules of the Road will be modified.

Summary of MSL Science Team Operations

After four Mars years and longer than seven Earth years on Mars, the Curiosity rover has been outstandingly successful in expanding our understanding of Mars as well as of Gale crater. The design, operations, and performance of the Science Team in operations, analysis of results, and publication have justified its ambitious goals. New goals are likely to be realized in the remaining years of operation.

The MSL Mission and Science Investigation, and its supplement, the MSL Science Team Rules of the Road, served well to prepare the members of the MSL Science Team to carry out their responsibilities, as did the ongoing guidance.

[CB Change] Coronavirus Pandemic The 90-day period when the operations teams all met at JPL ended in December, 2012, so in 2013, many members returned to their normal institutions. Since then, the operations teams worked with many members linked electronically. Since mid-March, 2020, all JPL Curiosity operations team members teleworked, after a short period for some hardware and software modifications that were needed to accomplish this, especially for navigation.

All JPL workers that can do so are now teleworking, including for strategic, supratactical, and tactical operations. The only activities that require presence at JPL are performed with recommended precautions. An example of such an activity that might be associated with the MSL Curiosity mission would be testing of a workaround procedure to overcome a problem at Mars.

12 Mars 2020: In the Tracks of Giants

Introduction

What's next? Curiosity's results lead to more questions, and to address them NASA launched another mission in summer, 2020. That mission, called simply "Mars 2020" builds on MSL's experience to change, enhance, and improve the mission and the rover. The Mars 2020 rover is named "Perseverance", and it carries a helicopter named "Ingenuity."

First, Mars 2020 will reuse much of the basic architecture and design work that MSL invented. For example, the TDS (Terminal Descent Sensor) landing radar software that MSL developed from scratch was simply copied, and will be reused exactly as it was.

Second, a different type of landing site was chosen, enabled by enhanced landing technology.

Third, the suite of science instruments is changed up. Some are upgrades of MSL instruments with new technology. Others will not be re-flown, having done their job.

Finally, some new instruments have been designed to answer new questions, expanding our understanding of the Red Planet in brand new ways. And additional experiments aim squarely at preparing for human exploration.

EDL (Entry, Descent, and Landing) Technology

In every Mars landing to date, we could only aim at a large landing area, the "landing ellipse," preselected to be "safe enough," and not to be unlucky enough to hit one of the few known hazards that any large area holds. Hazards include large boulders, or steep crater slopes. The Mars 2020 EDL design adds two great improvements.

1. The landing ellipse is 50% smaller, with the addition of range trigger technology to finely tune parachute-deployment timing. MSL was constrained to select 25 × 20 km landing areas; now, many more (smaller) areas are targetable.
2. TRN (Terrain Relative Navigation: see Fig. 12.1) is the new ability to determine in the last minutes that the spacecraft is heading unluckily toward a known hazard, and to divert to a safe spot. This is exactly what Apollo 11 did, on manual control, to avoid a boulder field.

This chapter is contributed by DJ Byrne as a private venture and not in his capacity as an employee of the Jet Propulsion Laboratory, California Institute of Technology.

C. J. Byrne, *Travels with Curiosity*, https://doi.org/10.1007/978-3-030-53805-7_12

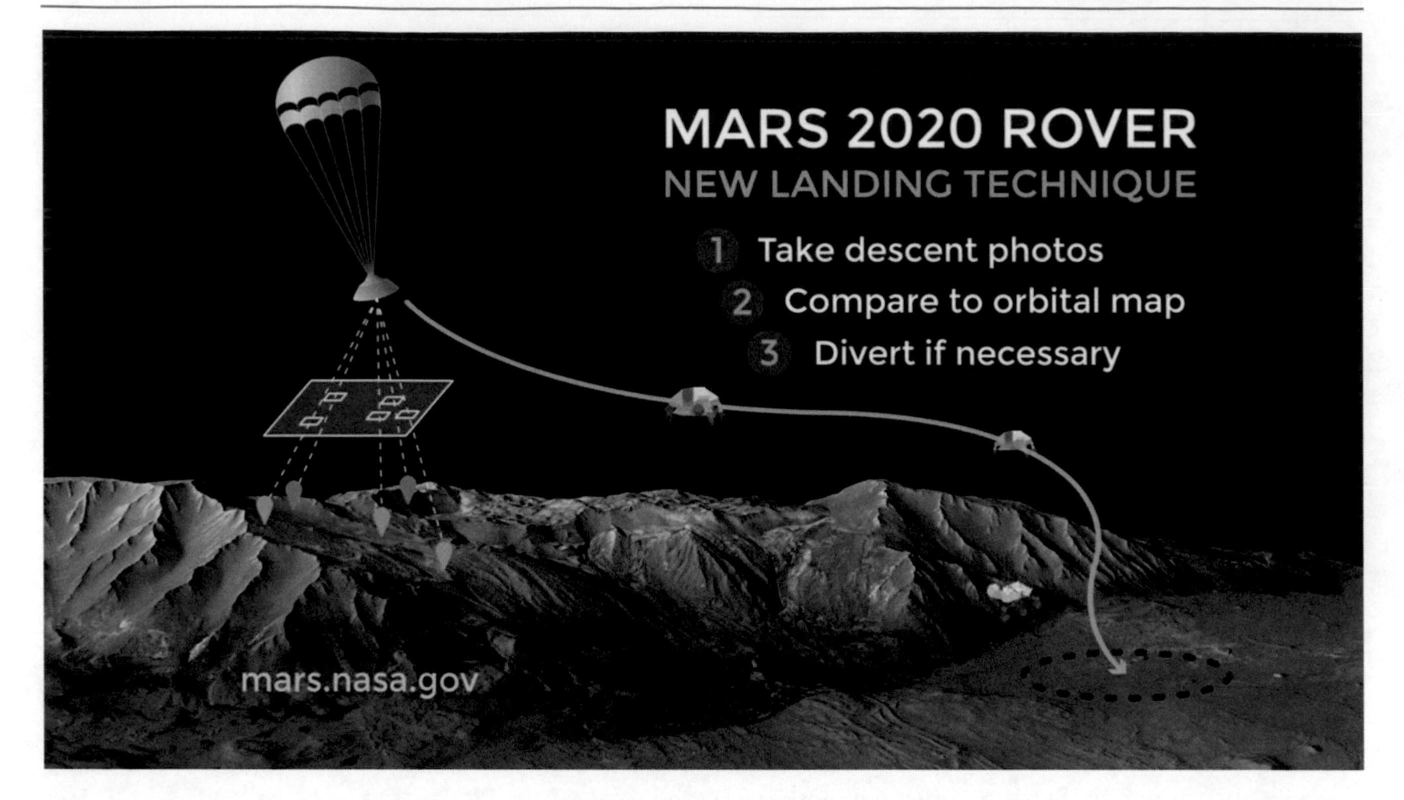

Fig. 12.1 Terrain relative navigation. (Image courtesy of NASA)

EDL Cameras Mars 2020 adds a suite of cameras to photograph, for the first time, the dramatic series of high-speed activities that occur far outside our view. Never before have we witnessed a supersonic parachute opening in the Martian atmosphere, the rover being lowered on a tether from its hovering descent stage to the surface, the tether being cut, or the descent stage flying away after rover touchdown. This will be invaluable to future missions improving on the technology. The images should also comprise a fantastic public engagement tool.

Landing Site Imagine studying Earth from only six locations! And all six in "safe to land" areas—say, the Arabian, Gobi, Saharan, Mojave, Atacama, and Tanami Deserts. The EDL technology improvements described above enabled selecting a more challenging but interesting destination. Jezero crater (see Fig. 12.2) includes an ancient lake-delta system, where a variety of rock types have been mixed together over time.

Thinking While Driving The MSL rover can drive itself for short distances by taking pictures and analyzing them on board to find a safe driving path. The analysis takes a long time, and the rover has to stop moving while it computes its next steps. Mars 2020 adds a "FastTraverse" ability to greatly speed up that processing, meaning that the rover need not pause its driving to continue in safety.

You might be thinking about self-driving cars moving along at highway speeds and wondering what's the big deal. It's actually quite impressive when you consider that the rover's computer is far, far less powerful than a car computer, because we traded off computing power for durability to help it survive and work in the vacuum of outer space on the long trip to Mars. Also, self-driving cars have the benefit of paved roads, painted lines, and GPS (Global Positioning System) satellite signals; they would not survive the unmarked terrain of Mars.

Fig. 12.2 Jezero crater. The western rim of the crater is shadowed to the left of center in this false color topographic map. (Image courtesy of NASA/JPL/JHUAPL/MSSS/Brown University)

Wheel Redesign Early in Curiosity's surface mobility, route planners avoided slippery sand and stayed on solid, rocky ground to keep firm footing and make the best progress. Yet over time, the aluminum wheels accumulated much more damage than expected. And not just a few spots thinning out—large holes became ripped out. Fig. 12.3 shows pictures of the same wheel surfaces taken several weeks apart. Note the gaping holes appearing over time, as though punched out with a juice can opener. That's not good; it would be difficult to rove without wheels. Nothing in the planning or test campaign predicted this level of wear so soon in the mission. It had not been observed on previous rovers.

So first, the drivers switched to preferring soft sand over hard pointy rocks. Second, the engineers mounted a fierce effort to reproduce the damage with test wheels on Earth, to understand what was causing it. And third, the operations team re-planned the overall route to avoid the kind of ground now known to be hazardous.

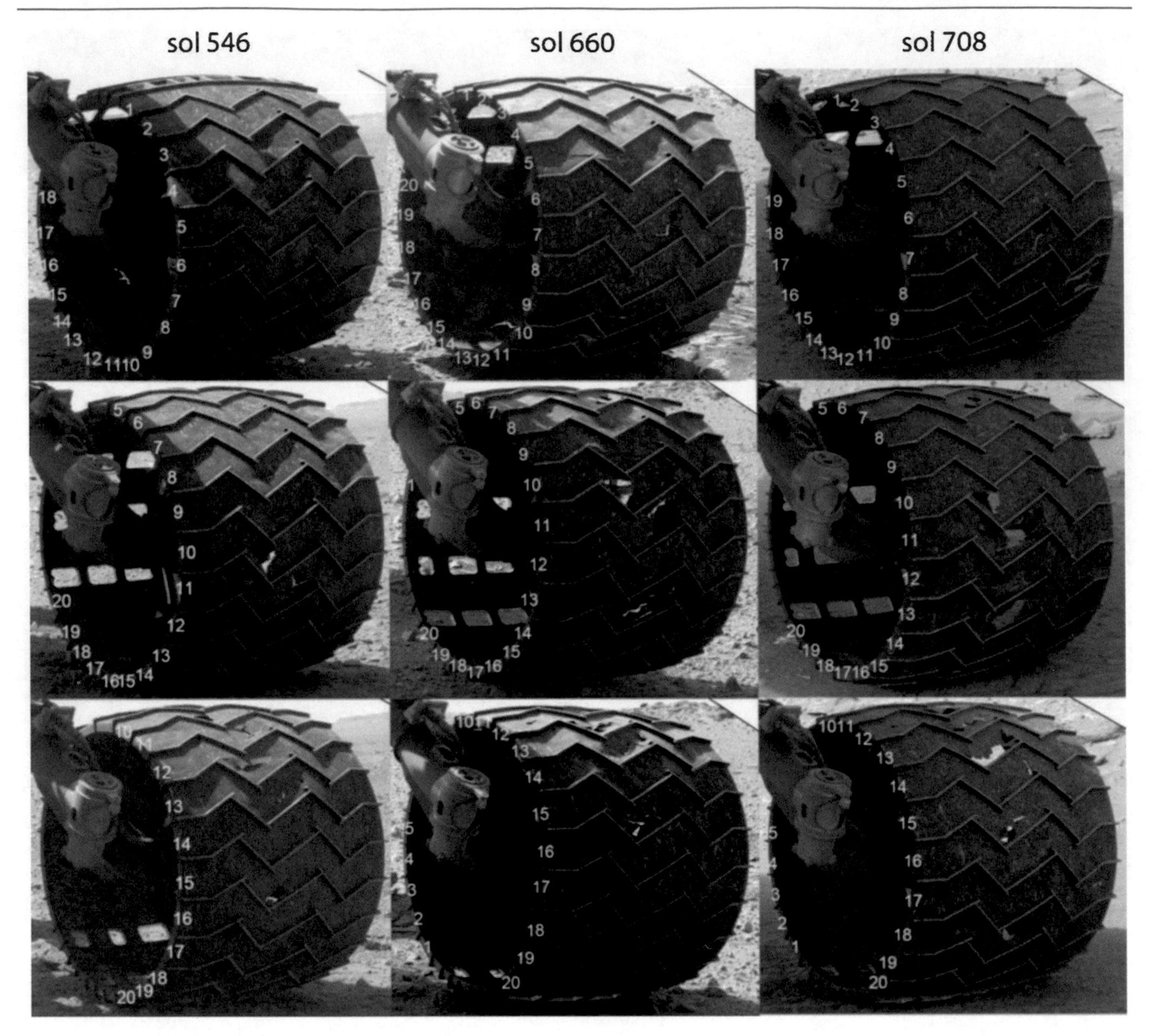

Fig. 12.3 Curiosity wheel damage over time and implied distance. (Image courtesy of NASA/JPL/MAHLI/Emily Lakdawalla)

In direct response to this harrowing experience by Curiosity, Perseverance will sport thicker, more durable aluminum wheels, with reduced width and a greater diameter (52.5 cm) compared to MSL's (50 cm) wheels.

Coring Drill and SCS (Sample Caching System) Both MSL and Mars 2020 carry drills to get at material from the interior of rocks. MSL's *percussive impact* drill grinds the rock to powder and gathers it up as it goes, and then delivers that powder to instruments in the rover body for analysis. Afterwards, the powder is dumped out to clear the way for the next sample.

We can learn much more about, and from, the rocks of Mars if we can bring samples home. Laboratories on Earth of course have a greater variety of more powerful instruments than any rover could possibly carry.

Mars 2020 includes a *coring* drill, which is hollow, sculpting out cylinders of rock to place in sample tubes for a still later potential mission to retrieve and bring to Earth. It will be extremely important, and tricky, to keep these few dozen samples pristine, so that they represent Mars and not contamination from Earth picked up along the way.

The Mars 2020 SCS is a complex set of robotic components to accept those possibly broken rock cylinders from the coring drill, seal them inside rugged tubes, and store the tubes inside the rover. Over many months, individual samples will be collected at differing sites as the rover explores Jezero crater. Eventually one or more depot locations will be chosen where the samples will be disgorged and left on the surface, so a later mission could pick them up from one place without repeating the lengthy exploration.

Mars 2020 Science Instrument Changes

Many of the instruments on the Perseverance rover have changed due to operational experience, new specific goals and advances in technology (see Fig. 12.4).

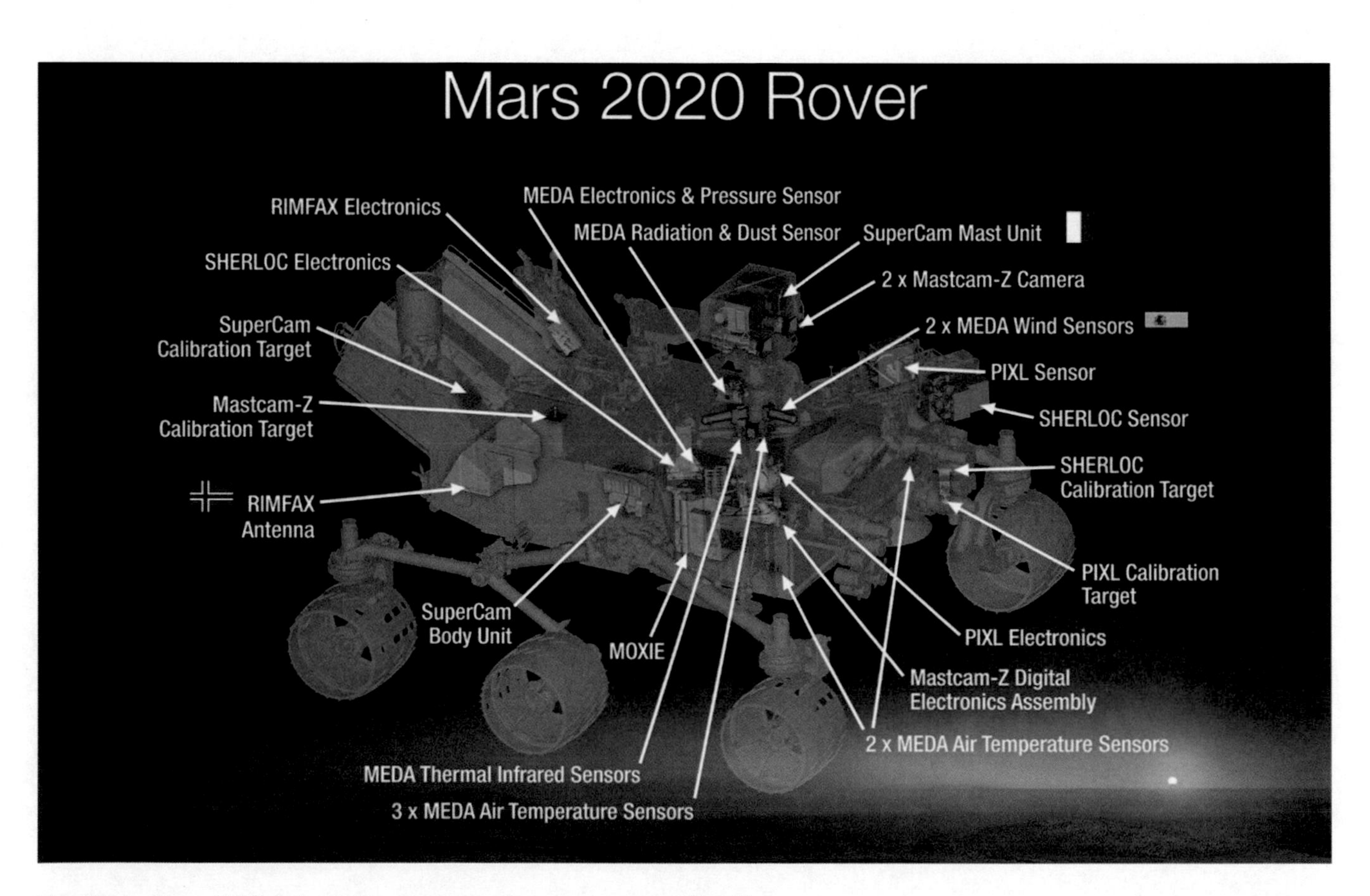

Fig. 12.4 Mars 2020 Science Instruments. (Image courtesy NASA)

Mars Environmental Dynamics Analyzer (MEDA) MSL's REMS (Rover Environmental Monitoring Station) instrument passes its legacy, via the Insight Mission's TWINS (Temperature and Wind for InSight) package, on to Mars 2020's MEDA. MEDA is a set of sensors that measure temperature, wind speed and direction, pressure, relative humidity, radiation, and dust size and shape.

One of MSL's lessons learned is how much dust and debris are kicked up in the final seconds of landing. The REMS wind sensor, we believe, was damaged at touchdown by that grit-storm and its results have been degraded ever since. MEDA's design is therefore modified based on our better understanding of the landing environment.

MastCam-Z Mars 2020 adds a zoom lens to the MSL Mastcam, and otherwise uses this excellent camera design as-is. The zoom lens allows us to gather higher resolution pictures around a site without time-consuming driving in all directions.

SuperCam MSL's ChemCam (Chemistry and Camera instrument) vaporizes tiny samples of rocks, analyzing the resulting plasma blast to determine what elements were in the sample. A black-and-white camera provides context for the sample used. Mars 2020 upgrades the black-and-white camera to color, and additional lasers and spectrometers allow analysis not just of elements, but of minerals and molecules.

PIXL (Planetary Instrument for X-ray Lithochemistry) MSL's CheMin performs X-ray spectroscopy on ground-up samples of rock powder. That is, the drill bores into a rock, and delivers the dust to CheMin, which is mounted inside the rover body.

On Mars 2020, PIXL is mounted at the end of the very long arm, analyzing minerals by scanning across a rock, thus gaining knowledge not only of what is there, but where different minerals are relative to each other.

SHERLOC (Scanning Habitable Environments with Raman and Luminescence for Organics and Chemicals) and WATSON (Wide Angle Topographic Sensor for Operations and Engineering) MSL's MAHLI (Mars Hand Lens Imager) self-focusing color camera, with white-light and ultraviolet light sources, examines fine detail on surface rocks. Mars 2020 adds to that spectrometers and a laser to gather a wider breadth of information. Also, SHERLOC carries small pieces of spacesuit material to examine over time, gathering data on how they hold up in the harsh Martian environment—valuable things to know for future astronauts!

New Science Instruments

(New!) RIMFAX (Radar Imager for Mars' Subsurface Experiment) You can't judge a book by its cover, and you can't understand a planet by just its surface. Ground-penetrating radar is a way to look beneath the surface and see how multiple layers of material, whether rock, sand, ice, or water, etc. lie on top of each other. NASA's Mars Reconnaissance Orbiter flies the SHARAD (Shallow Radar) instrument to see such changes over very large areas.

As Perseverance roves across the surface, its RIMFAX will peer down through the ground beneath it. RIMFAX can detect ice, water or salty brines of more than 30 feet (10 meters) beneath the surface of Mars.

(New!) MOXIE (Mars Oxygen In-Situ Resource Utilization Experiment) Humans are fond of breathing, and to do that you need O_2 (oxygen molecules). Sadly, Mars has less than a fraction of a percent as much O_2 in its atmosphere than Earth has. This has not been a problem so far, as all missions to Mars have been with robots.

Future missions landing astronauts on Mars may fill the need by breaking apart CO_2 (carbon-dioxide) molecules from the Martian air into plain C (carbon) and the needed O_2. Perseverance's MOXIE will be the first demonstration of this life-preserving technology on another planet. We need to know how much O_2 we can generate, and how much that varies depending on the air temperature, pressure, and how clear or dusty the incoming air is. MOXIE will characterize how well the process works over day/night cycles, over seasons, and during dust-storms.

Because there are no humans on board this mission to use the O_2 generated, MOXIE will simply vent it back outside. Those future missions will need to build on MOXIE's experience to figure out how much O_2 they can make, and how fast, at what energy cost. They will also need to work out how to store it.

(New!) "Ingenuity", the Mars Helicopter Scout The first heavier-than-air vehicle to fly on Mars! Quoting the NASA website, "The Mars Helicopter is considered a high-risk, high-reward technology demonstration. If the small craft encounters difficulties, the science-gathering of the Mars 2020 mission won't be impacted. If the helicopter does take flight as designed, future Mars missions could enlist second-generation helicopters to add an aerial dimension to their explorations."

Summary

The MSL and its rover, Curiosity have been and remain a fantastic mission—a technological marvel, and a scientific bonanza. She has taught us much about Mars, and by comparison, about our own world. The Mars 2020 rover, helicopter, and its new Entry, Descent, and Landing capability seek to be a worthy successor; capitalizing on rather than repeating what MSL has accomplished. Where similar observations of a new location make sense, the instruments have been improved. Where MSL observations prompted new questions, new types of instruments have been created.

Despite all the advancements, the instruments we can miniaturize and make work on a distant planet cannot compare to what can be accomplished with larger, more power-hungry instruments here at home. For example, while MSL and Mars 2020 each can tell us something about a rock's elements and molecules, neither can tell us what isotopes of potassium are present in what ratios. On Earth, we could do an isotope analysis, measuring not only what elements are present but how many neutrons (sub-atomic particles) they have, and that serves as a reliable clock for how long the rock material has been decaying since it was formed. Also, on Earth we do much finer microscopic analyses of interior grain structures from extremely thinly sliced rock samples, which helps us understand the conditions (wet vs. dry, under how much pressure, etc.) under which they were formed.

Mars 2020 will prepare a diverse cache for a planned future sample-return mission to bring to Earth, forging the next link in a chain of exploration.

Coronavirus and Remote Work

What does it mean to "work remotely?"

From the moment a Mars rover launches from Earth, your workplace is already as remote as it gets! But being remote from your team is something else again. The hundreds of MSL scientists working together while physically distanced from each other succeed through coordination with a much smaller group based at JPL. That small team is usually all in a single room, coordinating and distilling multiple conversations into a single set of commands for the rover. On March 20, 2020, in response to the Coronavirus outbreak, for the first time none of the team was present at JPL. They say it's taken some getting used to. Each day's planning takes longer than before to get the job done. This first all-remote activity drilled a rock sample at the "Edinburgh" location on sol 2713.

Once again, Curiosity has led the way that Perseverance is to follow. Yet the challenges are different. The shift to teleworking came just four months before Perseverance's July launch—a high-pressure time in the best of circumstances. The planets have to be aligned—literally—for launch: we get a three-week window every 26 months. To make that deadline takes a lot of hands-on work, both with the flight hardware at the launch site in Florida, and with the test-model copies at JPL in California.

Spacecraft components (launch vehicle, cruise stage, descent stage, and rover) were still arriving at the KSC (Kennedy Space Center) launch site as the team converted over to 90% teleworking. The well-named Perseverance mission became the highest priority within NASA, without compromising personnel safety. Those performing mission-essential on-site work were already well-trained and accustomed to wearing masks and other gear to protect the hardware; they updated the procedures to protect each other from virus and kept on working. As of this writing, the spacecraft has launched and is on target to land on February18, 2021.

How would the Mars 2020 launch have been affected if the pandemic had hit a year earlier, when there was still flight hardware being assembled and tested at JPL? Impossible to say. But the team would have wanted to persevere.

I would like to thank the NASA outreach folks, the Planetary Society, and Wikipedia for making all of the great information above publicly available. The material in this chapter is drawn from the following public websites:

https://mars.nasa.gov/

https://science.nasa.gov/technology/technology-stories/Computing-Advances-Enable-New-Rover-Red-Planet

https://www.nasa.gov/press-release/nasa-announces-landing-site-for-mars-2020-rover

https://www.nasa.gov/feature/jpl/a-neil-armstrong-for-mars-landing-the-mars-2020-rover

https://pds.nasa.gov/ds-view/pds/viewProfile.jsp?dsid=MSL-M-REMS-4-ENVRDR-V1.0

https://mepag.jpl.nasa.gov/reports/iMOST_Final_Report_180814.pdf

https://nasapeople.nasa.gov/coronavirus/

https://www.planetary.org/blogs/emily-lakdawalla/2014/08190630-curiosity-wheel-damage.html

https://en.wikipedia.org/wiki/Mars_2020

In addition, there is material from "Heavyweight Quality, Agile Methods" by DJ Byrne, the keynote presentation for the Agility in Flight Mini-Workshop, 2019 IEEE International Conference on Space Mission Challenges for Information Technology, Pasadena, California.

Appendix: The MSL Science Team

Introduction: The MSL Science Team is the group of scientists from American and International institutions who proposed and designed the instruments that Curiosity carries to investigate Gale crater. The team includes those who manage the operations, staff the operations groups, and analyze and archive the returned data. They also interpret the results of the investigation and publish the results.

There are two groups of scientists in the MSL Science Team: Science Team Members and Science Team Collaborators.

Science Team Members are the scientists who will manage the operations of Curiosity, making decisions strategically and tactically on a day to day basis. Representatives will meet each sol (the Martian Day) to plan the detailed operations of each short period in response to the data returned from the previous period. At times they plan 2 or 3 sols at once to take time off, usually for a weekend. This group was physically present at JPL for the first 90 sols after Curiosity landed at Gale crater. After that, members returned to their normal institutions and participated by phone.

Science Team Collaborators support the planning groups as they are needed, and participate in the other functions of the MSL Science Team as well.

The personnel structure of the Mars Science Laboratory Science Team was established by a document entitled "Mars Science Laboratory Science Team Rules of the Road," a supplement to a peer-reviewed paper entitled "MSL Mission and Science Team Investigation." Both documents are described (with notes on their application during Curiosity's exploration) in Chap. 11 of this book.

The personnel of the initial members of the MSL Science Team are listed below in association with their science instrument and the Principal Investigator in section "Science Team Members" of this appendix. The science team collaborators are listed below with their science instrument, if applicable, in section "Science Team Collaborators" of this appendix. Each person is identified as a Principal Investigator or Co-investigator as applicable, and their host institution is named. The list is an impressive roster of about 400 experts in planetary science and instrumentstion from a diverse set of institutions, both American and International

Personnel Lists

Science Team Members

The following individuals are the members of the MSL Science Team[1]:

MSL Project Science Office + NASA Headquarters
Anderson, Robert C., JPL (SA/SPaH Investigation Scientist)

[1]Names in bold indicate Project Science Group (PSG) members.

C. J. Byrne, *Travels with Curiosity*, https://doi.org/10.1007/978-3-030-53805-7

Behar, Alberto, JPL (DAN Investigation Scientist)
Blaney, Diana, JPL (ChemCam Investigation Scientist)
Brinza, David, JPL (RAD Investigation Scientist)
Crisp, Joy, JPL (MSL Deputy Project Scientist)
de la Torre Juarez, Manuel, JPL (REMS Investigation Scientist)
Grotzinger, John, Caltech (MSL Project Scientist)
Maki, Justin, JPL (MastCam, MAHLI, MARDI Investigation Scientist)
Meyer, Michael, NASA Headquarters (Program Scientist)
Vasavada, Ashwin, JPL (MSL Deputy Project Scientist)
Voytek, Mary, NASA Headquarters (Deputy Program Scientist)
Yen, Albert, JPL (CheMin and APXS Investigation Scientist)

APXS
Campbell, Iain, University of Guelph (Co-I)
Gellert, Ralf, University of Guelph (PI)
King, Penny, Australia National University (Co-I)
Leshin, Laurie Rensselaer, Polytechnic Institute (RPI) (Co-I)
Lugmair, Guenter, University California San Diego (Co-I)
Spray, John, University New Brunswick (Co-I)
Squyres, Steven, Cornell University (Co-I)
Yen, Albert, JPL (Co-I)

ChemCam
Blaney, Diana, JPL (Co-I)
Bridges, Nathan, Applied Physics Laboratory (APL) (Co-I)
Clark, Benton, Space Science Inst. (SSI) (Co-I)
Clegg, Sam, Los Alamos National Lab (LANL) (Co-I)
Cremers, David, Applied Research Associates, Inc. (Co-I)
Dromart, Gilles, Laboratoire de Géologie de Lyon (LGLTPE) (Co-I)
d'Uston, Claude, Institut de Recherche en Astrophysique et Planetologie (IRAP) (Co-I)
Fabre, Cécile, Géologie et Gestion des Ressources Minérales et Energétiques (G2R) (Co-I)
Gasnault, Olivier, Institut de Recherche en Astrophysique et Planetologie (IRAP) (Co-I)
Herkenhoff, Ken, U.S. Geological Survey (USGS) Flagstaff (Co-I)
Kirkland, Laurel, LPI (Co-I)
Langevin, Yves, Institut d'Astrophysique Spatiale (IAS) (Co-I)
Mangold, Nicolas, Laboratoire de Planétologie et Géodynamique de Nantes (LPGN) (Co-I)
Mauchien, Patrick, Commissariat à l'Énergie Atomique et aux Énergies Alternatives (CEA) (Co-I)
Maurice, Sylvestre, Institut de Recherche en Astrophysique et Planetologie (IRAP) (Co-I and Deputy PI)
McKay, Christopher, Ames Research Center (Co-I)
Newsom, Horton, University New Mexico (Co-I)
Sautter, Violaine, Laboratoire de Minéralogie et Cosmochimie du Muséum (LMCM) (Co-I)
Vaniman, David, Planetary Science Institute (PSI) (Co-I)
Wiens, Roger Craig, Los Alamos National Lab (LANL) (PI)

CheMin

Anderson, Robert, JPL (Co-I)
Bish, David, Indiana University Bloomington (Co-I)
Blake, David, F. Ames Research Center (PI)
Chipera, Steve, Chesapeake Energy (Co-I)
Crisp, Joy, JPL (Co-I)
DesMarais, David, Ames Research Center (Co-I)
Downs, Bob, University Arizona (Co-I)
Farmer, Jack, Arizona State University (ASU) (Co-I)
Feldman, Sabrina, JPL (Co-I)
Gailhanou, Marc, Centre National de la Recherche Scientifique (CNRS) (Co-I)
Ming, Douglas, JSC (Co-I)
Morris, Richard, JSC (Co-I)
Sarrazin, Philippe, inXitu (Co-I)
Stolper, Ed, Caltech (Co-I)
Treiman, Allan, Lunar and Planetary Institute (LPI) (Co-I)
Vaniman, David, Planetary Science Institute (PSI) (Co-I and Deputy PI)
Yen, Albert, JPL (Co-I)

DAN

Behar, Alberto, JPL (Co-I)
Boynton, Bill, University Arizona (Co-I)
Kozyrev, Alexandre S., Space Research Inst. (IKI) (Co-I)
Litvak, Maxim, Space Research Inst. (IKI) (Co-I and Deputy PI)
Mitrofanov, Igor G., Space Research Inst. (IKI) (PI)
Sanin, Anton B., Space Research Inst. (IKI) (Co-I)

MAHLI, MARDI, and MastCam

Bell, James F. III, Arizona State University (Co-I)
Cameron, James, Lightstorm Entertainment Inc. (Co-I)
Dietrich, William E., University California Berkeley (Co-I)
Edgett, Kenneth S., Malin Space Science Systems (MSSS) (MAHLI PI)
Edwards, Laurence, Ames Research Center (Co-I)
Garvin, James B., GSFC (Co-I)
Hallet, Bernard, University of Washington, Seattle (Co-I)
Herkenhoff, Kenneth E., U.S. Geological Survey (USGS) Flagstaff (Co-I)
Heydari, Ezat, Jackson State University (Co-I)
Kah, Linda C., University Tennessee Knoxville (Co-I)
Lemmon, Mark T., Texas A&M (Co-I)
Maki, Justin, JPL (Co-I)
Malin, Michael C., Malin Space Science Systems (MSSS) (MastCam & MARDI PI)
Minitti, Michelle E., Arizona State University (Co-I)
Olson, Timothy S., Salish Kootenai College (Co-I)
Parker, Timothy J., JPL (Co-I)
Rowland, Scott K., University of Hawaii Manoa (Co-I)
Schieber, Juergen, Indiana University Bloomington (Co-I)

Sullivan, Robert J., Cornell University (Co-I)
Sumner, Dawn Y., University California Davis (Co-I)
Thomas, Peter C., Cornell University (Co-I)
Yingst, Aileen R., Planetary Science Institute (PSI) (at University Wisconsin Green Bay) (Co-I and MAHLI Deputy PI)

RAD

Boehm, Eckart, University Kiel (Co-I)
Böttcher, Stephan, University Kiel (Co-I)
Brinza, David, JPL (Co-I)
Bullock, Mark, Southwest Research Institute (SwRI) (Co-I)
Burmeister, Sonke, University Kiel (Co-I)
Cleghorn, Timothy, JSC (Co-I)
Cucinotta, Frank, JSC (Co-I)
Grinspoon, David, Denver Museum of Nature & Science (Co-I)
Hassler, Donald, Southwest Research Institute (SwRI) (PI)
Martín García, César, University Kiel (Co-I)
Mueller-Mellin, Reinhold, University Kiel (Co-I)
Posner, Arik, NASA Headquarters (Co-I)
Rafkin, Scot, Southwest Research Institute (SwRI) (Co-I)
Reitz, Günther, Deutsches Zentrum für Luft- und Raumfahrt (DLR) (Co-I)
Wimmer-Schweingruber, Robert, University Kiel (Co-I)
Zeitlin, Cary, Southwest Research Institute (SwRI) (Co-I)

REMS

Gómez-Elvira, Javier, Centro de Astrobiología (CAB-CSIC/INTA) (PI)
Haberle, Robert, Ames Research Center (Co-I)
Gómez Gómez, Felipe, Centro de Astrobiología (CAB-CSIC/INTA) (Co-I)
Harri, Ari-Matti, Finnish Meteorological Institute (FMI) (Co-I)
Martínez-Frías, Jesús, Centro de Astrobiología (CAB-CSIC/INTA) (Co-I)
Martín-Torres, F. Javier, Centro de Astrobiología (CAB-CSIC/INTA) (Co-I)
Ramos, Miguel, Universidad de Alcalá de Henares (Co-I)
Renno, Nilton, University Michigan Ann Arbor (Co-I)
Richardson, Mark, Ashima Research (Co-I)
Rodriguez Manfredi, José, Antonio Centro de Astrobiología (CAB-CSIC/INTA) (Co-I)
Sebastian Martinez, Eduardo, Centro de Astrobiología (CAB-CSIC/INTA) (Co-I)
Torre Juarez, Manuel, JPL (Co-I)
Zorzano Mier, María-Paz, Centro de Astrobiología (CAB-CSIC/INTA) (Co-I)

SAM

Atreya, Sushil, University Michigan Ann Arbor (Co-I)
Brinckerhoff, William, GSFC (Co-I)
Cabane, Michel, Laboratoire Atmospheres, Milieux, Observations Spatiales (LATMOS) (Co-I)
Coll, Patrice, Laboratoire Interuniversitaire des Systemses Atmospheriques (LISA) (Co-I)
Conrad, Pamela, GSFC (Co-I and Deputy PI)
Goesmann, Fred, Max Planck Institute for Solar System Research (Co-I)
Gorevan, Stephen, Honeybee Robotics (Co-I)

Jakosky, Bruce, University Colorado Boulder (Co-I)
Jones, John, JSC (Co-I)
Leshin, Laurie, Rensselaer Polytechnic Institute (RPI) (Co-I)
Mahaffy, Paul, GSFC (PI)
McKay, Christopher, Ames Research Center (Co-I)
Ming, Douglas, JSC (Co-I)
Morris, Richard, JSC (Co-I)
Navarro-González, Rafael, University Nacional Autónoma de México (Co-I)
Owen, Tobias, University Hawaii at Manoa (Co-I)
Pepin, Robert, University Minnesota (Co-I)
Raulin, François, Laboratoire Interuniversitaire des Systemses Atmospheriques (LISA) (Co-I)
Robert, François, Laboratoire de Minéralogie et Cosmochimie du Muséum (LMCM) (Co-I)
Squyres, Steve,n Cornell University (Co-I)
Steele, Andrew, Carnegie Inst. Washington (Co-I)
Webster, Chris, JPL (Co-I)

Participating Scientists
Arvidson, Ray, Washington University St. Louis
Bridges, John, University of Leicester
Dyar, Darby, Mt. Holyoke College
Ehlmann, Bethany, Caltech
Eigenbrode, Jen, GSFC
Farley, Ken, Caltech
Fisk, Marty, Oregon State University
Glavin, Daniel, GSFC
Goetz, Walter, Max Planck Institute for Solar System Research
Grant, John, Smithsonian Institution
Gupta, Sanjeev, Imperial College of Science, Technology and Medicine
Hamilton, Vicky, Southwest Research Institute (SwRI)
Johnson, Jeffrey, Applied Physics Laboratory (APL)
Jun, Insoo, JPL
Kocurek, Gary, University of Texas at Austin
Léveillé, Richard, Canadian Space Agency (Canadian Space Agency (CSA)
Lewis, Kevin, Princeton University
Madsen, Morten, University of Copenhagen
McLennan, Scott, SUNY Stony Brook
Milliken, Ralph, Brown University
Mischna, Michael, JPL
Moersch, Jeff, University Tennessee Knoxville
Moores, John, Western University, Ontario
Niles, Paul, JSC
Oehler, Dorothy, JSC
Rubin, David, U.S. Geological Survey (USGS) Santa Cruz
Schmidt, Mariek, Brock University
Smith, Michael, GSFC
Summons, Roger, MIT
Williams, Rebecca, Planetary Science Institute (PSI)

Science Team Collaborators

The following collaborators are associated with MSL science team members:

MSL Project Science Office

Beegle, Luther, JPL
Calef, Fred, JPL
DeFlores, Lauren, JPL
Edgar, Lauren, Caltech (student of John Grotzinger)
Feldman, Jason, JPL
Griffes, Jennifer, Caltech
Hurowitz, Joel, JPL
Milkovich, Sarah, JPL
Morookian, John Michael, JPL
Pavri, Betina, JPL
Rice, Melissa, Caltech (NASA Postdoc Program)
Scodary, Anthony, JPL
Sengstacken, Aaron, JPL
Siebach, Kirsten, Caltech (student of John Grotzinger)
Simmonds, Jeff, JPL
Spanovich, Nicole, JPL
Stack, Katie, Caltech (student of John Grotzinger)

APXS

Berger, Jeffery, University New Mexico (student of Penny King and Horton Newsom)
Boyd, Nick, University of Guelph
Brunet, Claude, Canadian Space Agency (CSA)
Elliott, Beverley, University New Brunswick
Hipkin, Victoria, Canadian Space Agency (CSA)
Marchand, Geneviève, Canadian Space Agency (CSA)
Perrett, Glynis, University of Guelph (student of Ralf Gellert)
Pradler, Irina, University of Guelph
Thompson, Lucy, University New Brunswick
VanBommel, Scott, University of Guelph (student of Iain Campbell and Ralf Gellert)

ChemCam

Agard, Christophe, Centre National d'Etudes Spatiales (CNES)
Anderson, Ryan, U.S. Geological Survey (USGS) Flagstaff
Baratoux, David, Institut de Recherche en Astrophysique et Planetologie (IRAP)
Baroukh, Julien, Centre National d'Etudes Spatiales (CNES)
Barraclough, Bruce, Planetary Science Institute (PSI)
Bender, Steve, Planetary Science Institute (PSI)
Berger, Gilles, Institut de Recherche en Astrophysique et Planetologie (IRAP)
Blank, Jen, Bay Area Environment Research Institute (BAER) (at Ames Research Center)
Charpentier, Antoine, ATOS Origin
Cousin, Agnès, Institut de Recherche en Astrophysique et Planetologie (IRAP)
Cros, Alain, Institut de Recherche en Astrophysique et Planetologie (IRAP)

DeLapp, Dorothea, Los Alamos National Lab (LANL)
Dingler, Robert, Los Alamos National Lab (LANL)
Donny, Christophe, Centre National d'Etudes Spatiales (CNES)
Dupont, Audrey, CS Systemes d'Information
Favot, Laurent, Capgemini France
Forni, Olivier, Institut de Recherche en Astrophysique et Planetologie (IRAP)
Gaboriaud, Alain, Centre National d'Etudes Spatiales (CNES)
Gondet, Brigitte, Institut d'Astrophysique Spatiale
Guillemot, Philippe, Centre National d'Etudes Spatiales (CNES)
Johnstone, Steve, Los Alamos National Lab (LANL)
Lacour, Jean Luc, Commissariat à l'Énergie Atomique et aux Énergies Alternatives (CEA)
Lafaille, Vivian, Centre National d'Etudes Spatiales (CNES)
Lanza, Nina, Los Alamos National Lab (LANL)
Lasue, Jeremie, Institut de Recherche en Astrophysique et Planetologie (IRAP)
Lee, Qiu-Mei, Institut de Recherche en Astrophysique et Planetologie (IRAP)
Le Mouélic, Stéphane, Laboratoire de Planétologie et Géodynamique de Nantes (LPGN)
Lewin, Éric, Institut des Sciences de la Terre (ISTerre)
Lorigny, Eric, Centre National d'Etudes Spatiales (CNES)
Lundberg, Angela, Delaware State University
Manhes, Gérard, Inst. Physique du Globe de Paris (IPGP)
Melikechi, Noureddine, Delaware State University
Meslin, Pierre-Yves, Institut de Recherche en Astrophysique et Planetologie (IRAP)
Mezzacappa, Alissa, Delaware State University (student of Noureddine Melikechi)
Nelson, Tony, Los Alamos National Lab (LANL)
Ollila, Ann, University New Mexico (student of Horton Newsom)
Paillet, Alexis, Centre National d'Etudes Spatiales (CNES)
Pallier, Etienne, Institut de Recherche en Astrophysique et Planetologie (IRAP)
Parot, Yann, Institut de Recherche en Astrophysique et Planetologie (IRAP)
Peret, Laurent, ATOS Origin
Perez, René, Centre National d'Etudes Spatiales (CNES)
Pinet, Patrick, Institut de Recherche en Astrophysique et Planetologie (IRAP)
Poitrasson, Franck, Géosciences Environnement Toulouse (GET)/Centre National de la Recherche Scientifique (CNRS)
Radziemski, Leon, Piezo Energy Technologies, Tucson
Saccoccio, Muriel, Centre National d'Etudes Spatiales (CNES)
Schröder, Susanne, Institut de Recherche en Astrophysique et Planetologie (IRAP)
Sirven, Jean-Baptiste, Commissariat à l'Énergie Atomique et aux Énergies Alternatives (CEA)
Thulliez, Emmanuel, Centre National d'Etudes Spatiales (CNES)
Tokar, Robert, Planetary Science Institute (PSI)
Toplis, Mike, Institut de Recherche en Astrophysique et Planetologie (IRAP)
Williams, Joshua, University of New Mexico (student of Horton Newsom)
Yana, Charles, Centre National d'Etudes Spatiales (CNES)

CheMin

Achilles, Cherie, Jacobs Technology (at JSC) (student of Dick Morris and Doug Ming)
Bristow, Thomas, Ames Research Center (NASA Postdoc Program)

Brunner, Will, inXitu
Hoehler, Tori, Ames Research Center
Morrison, Shaunna, University Arizona (student of Bob Downs)
Rampe, Elizabeth, JSC (NASA Postdoc Program)
Wilson, Mike, Ames Research Center

DAN

Fedosov, Fedor, Space Research Inst. (IKI)
Fitzgibbon, Mike, University Arizona
Golovin, Dmitry, Space Research Inst. (IKI)
Harshman, Karl, University Arizona
Karpushkina, Natalya, Space Research Inst. (IKI) (student of Maxim Litvak)
Malakhov, Alexey, Space Research Inst. (IKI)
Mokrousov, Maxim, Space Research Inst. (IKI)
Nikiforov, Sergey, Space Research Inst. (IKI) (student of Maxim Litvak)
Prochorov, Vasily, Space Research Inst. (IKI)
Shterts, Ruslan, Space Research Inst. (IKI) (student of Maxim Litvak)
Tretyakov, Vladislav, Space Research Inst. (IKI)
Varenikov, Alexey, Space Research Inst. (IKI)
Vostrukhin, Andrey, Space Research Inst. (IKI)

MAHLI, MARDI, and Mastcam

Bean, Keri, Texas A&M (student of Mark Lemmon) Baker, Burt Malin, Space Science Systems (MSSS)
Cantor, Bruce, Malin Space Science Systems (MSSS)
Caplinger, Michael, Malin Space Science Systems (MSSS)
Davis, Scott, Malin Space Science Systems (MSSS)
Duston, Brian, Malin Space Science Systems (MSSS)
Fay, Donald, Malin Space Science Systems (MSSS)
Flückiger, Lorenzo, Carnegie Mellon University (CMU) (at Ames Research Center)
Godber, Austin, Arizona State University
Hardgrove, Craig, Malin Space Science Systems (MSSS)
Harker, David, Malin Space Science Systems (MSSS)
Herrera, Paul, Malin Space Science Systems (MSSS)
Hudgins, Judy, Salish Kootenai College (student of Tim Olson)
Jensen, Elsa, Malin Space Science Systems (MSSS)
Keely, Leslie, Ames Research Center
Krezoski, Jill, Malin Space Science Systems (MSSS)
Krysak, Daniel, Malin Space Science Systems (MSSS)
Lees, David, Carnegie Mellon University (CMU) (at Ames Research Center)
Lipkaman, Leslie, Malin Space Science Systems (MSSS)
McCartney, Elaina, Malin Space Science Systems (MSSS)
McNair, Sean, Malin Space Science Systems (MSSS)
Nefian, Ara, Carnegie Mellon University (CMU) (at Ames Research Center)
Nixon, Brian, Malin Space Science Systems (MSSS)
Palucis, Marisa, University California Berkeley
Posiolova, Liliya, Malin Space Science Systems (MSSS)
Ravine, Michael, Malin Space Science Systems (MSSS)

Sandoval, Jennifer, Malin Space Science Systems (MSSS)
Sletten, Ronald, University Washington Seattle
Stewart, Noel, Salish Kootenai College (student of Tim Olson)
Sucharski, Bob, U.S. Geological Survey (USGS) Flagstaff
Supulver, Kimberley, Malin Space Science Systems (MSSS)
Van Beek, Jason, Malin Space Science Systems (MSSS)
Van Beek, Tessa, Malin Space Science Systems (MSSS)
Williams, Amy, University Calif. Davis (student of Dawn Sumner)
Wu, Megan, Malin Space Science Systems (MSSS)
Zimdar, Robert, Malin Space Science Systems (MSSS)

RAD

DeMarines, Julia, Denver Museum of Nature & Science
Ehresmann, Bent, Southwest Research Institute (SwRI)
Kim, Myung-Hee Y., USRA (at JSC)
Köhler, Jan, University Kiel
Kortmann, Onno, University California Berkeley
Peterson, Joe, Southwest Research Institute (SwRI)
Plante, Ianik, USRA (at JSC)
Weigle, Eddie, Southwest Research Institute (SwRI)

REMS

Alves Verdasca, José Alexandre, Centro de Astrobiología (CAB-CSIC/INTA)
Armiens-Aparicio, Carlos, Centro de Astrobiología (CAB-CSIC/INTA)
Braswell, Shaneen, University Michigan (student of Nilton Renno)
Blanco Avalos, Juan J., University Alcalá de Henares
Carrasco Blázquez, Isaías, Centro de Astrobiología (CAB-CSIC/INTA) (student of J.A. Rodríguez-Manfredi)
Elliott, Harvey, University Michigan Ann Arbor (student of Nilton Renno)
Genzer, Maria, Finnish Meteorological Institute (FMI)
Halleaux, Douglas, University Michigan Ann Arbor (student of Nilton Renno)
Kahanpää, Henrik, Finnish Meteorological Institute (FMI)
Kahre, Melinda, Ames Research Center
Kemppinen, Osku, Finnish Meteorological Institute (FMI)
Lepinette Malvitte, Alain, Centro de Astrobiología (CAB-CSIC/INTA)
Martín-Soler, Javier, Centro de Astrobiología (CAB-CSIC/INTA)
McEwan, Ian, Ashima Research
Mora-Sotomayor, Luis, Centro de Astrobiología (CAB-CSIC/INTA)
Muñoz Caro, Guillermo M., Centro de Astrobiología (CAB-CSIC/INTA)
Navarro López, Sara, Centro de Astrobiología (CAB-CSIC/INTA)
Newman, Claire, Ashima Research
Pablo Hernández, Miguel, Ángel de University Alcalá de Henares
Peinado-González, Verónica, Centro de Astrobiología (CAB-CSIC/INTA)
Polkko, Jouni, Finnish Meteorological Institute (FMI)
Romeral-Planello, Julio, Centro de Astrobiología (CAB-CSIC/INTA)
Torres Redondo, Josefina, Centro de Astrobiología (CAB-CSIC/INTA)
Urqui-O'Callaghan, Roser, Centro de Astrobiología (CAB-CSIC/INTA)

SAM

Archer, Doug, JSC (NASA Postdoc Program)
Benna, Mehdi, University Maryland Baltimore County (UMBC) (at GSFC)
Bleacher, Lora, USRA-LPI (at GSFC)
Botta, Oliver, Swiss Space Office (SSO)
Buch, Arnaud, Laboratoire de Génie des Procédés et Matériaux (LGPM), Ecole Centrale Paris
Coscia, David, Laboratoire Atmospheres, Milieux, Observations Spatiales (LATMOS)
Dworkin, Jason, GSFC
Eigenbrode, Jen, GSFC
Flesch, Greg, JPL
Franz, Heather, University Maryland Baltimore County (UMBC) (at GSFC)
Freissinet, Caroline, GSFC (NPP)
Geffroy, Claude, Institut de Chimie des Milieux et Matériaux de Poitiers (IC2MP)
Glavin, Daniel, GSFC
Harpold, Daniel, GSFC
Huntress, Wesley, Carnegie Inst. Washington
Israël, Guy, Laboratoire Atmospheres, Milieux, Observations Spatiales (LATMOS)
Jones, Andrea, USRA-LPI (at GSFC)
Kasprzak, Wayne, GSFC
Keymeulen, Didier, JPL
Lefavor, Matthew, Microtel (at GSFC)
Lorigny, Eric, Centre National d'Etudes Spatiales (CNES)
Lyness, Eric, Microtel (at GSFC)
Malespin, Charles, USRA (at GSFC)
Manning, Heidi, Concordia College
Martin, David, GSFC
McAdam, Amy, GSFC
Nealson, Kenneth, University Southern California
Noblet, Audrey, Laboratoire Interuniversitaire des Systemses Atmospheriques (LISA) (student of Patrice Coll)
Nolan, Thomas, Nolan Engineering (at GSFC)
Patel, Kiran, Global Science & Technology, Inc. (at GSFC)
Pavlov, Alex, GSFC
Prats, Benito, eINFORMe Inc. (at GSFC)
Raaen, Eric, GSFC
Stern, Jennifer, GSFC
Sutter, Brad, Jacobs Technology (at JSC)
Szopa, Cyril, Laboratoire Atmospheres, Milieux, Observations Spatiales (LATMOS)
Tan, Florence, GSFC
Teinturier, Samuel, Laboratoire Atmospheres, Milieux, Observations Spatiales (LATMOS)
Trainer, Melissa, GSFC
Vicenzi, Edward, Smithsonian Inst.
Wadhwa, Meenakshi, Arizona State University
Wong, Michael H., University Michigan Ann Arbor Wray, James, Georgia Tech.

Associated with Participating Scientists

Aubrey, Andrew, JPL
Bentz, Jennifer, University of Saskatchewan (student of Richard Léveillé)
Berlanga, Genesis, Mount Holyoke College (student of M. Darby Dyar)
Berman, Daniel, Planetary Science Institute (PSI)
Breves, Elly, Mount Holyoke College (student of M. Darby Dyar)
Carmosino, Marco, University of Massachusetts (at Mount Holyoke College, student of M. Darby Dyar)
Choi, David, GSFC (NASA Postdoc Program)
Cloutis, Ed, University of Winnipeg
Cody, George, Carnegie Institution of Washington
Day, Mackenzie, University of Texas, Austin (student of Gary Kocurek)
Ewing, Ryan, University of Alabama
Fassett, Caleb, Mount Holyoke College
Floyd, Melissa, NASA GSFC
Fraeman, Abigail, Washington University St. Louis (student of Ray Arvidson)
Francis, Raymond, Western University, Ontario (student of John Moores)
French, Katherine Louise, MIT (student of Roger Summons)
Hayes, Alexander, University of California Berkeley
Hviid, Stubbe, Max Planck Inst. for Solar System Research
Iagnemma, Karl, MIT
Martin, Mildred, Catholic University of America (at GSFC)
McConnochie, Timothy, University of Maryland (at NASA GSFC)
McCullough, Emily, Western University, Ontario (student of John Moores)
Miller, Hayden, Caltech (student of Ken Farley)
Miller, Kristen, MIT
Muller, Jan-Peter, University College London
Noe Dobrea, Eldar, PSI
Ozanne, Marie, Mount Holyoke College (student of M. Darby Dyar)
Popa, Radu, Portland State University
Purdy, Sharon, Wilson Smithsonian Institution (student of John Grant)
Scholes, Dan, Washington University St. Louis (PDS support)
Schwenzer, Susanne, Open University
Slavney, Susie, Washington University St. Louis (PDS support)
Sobrón Sánchez, Pablo, Canadian Space Agency
Stalport, Fabien, Laboratoire Interuniversitaire des Systemses Atmospheriques (LISA)
Stein, Thomas, Washington University St. Louis (PDS support)
Tate, Christopher, University Tennessee Knoxville (student of Jeff Moersch) ten Kate, Inge, Utrecht University Vasconcelos, Paulo University Queensland
Ward, Jen, Washington University St. Louis (PDS support)
Westall, Frances, CNRS
Wolff, Michael, Space Science Institute (SSI)

Glossary of Geology Terms

- Aeolian — A process of moving sand or other material by wind.
- Amorphous — A material without crystal structure, like mud or silt. A component of a sample that does not show a diffraction pattern when exposed to an X-ray beam.
- Bagnold Dunes — A type of dunes that have a specific form, whether on Earth or Mars. Ralph A. Bagnold, a geologist who explored the Libyan desert in 1932, wrote a book on the form of dunes. NASA named the Bagnold Dunes of Mars after him. He commanded the Long Range Desert Group in the British Army in World War II, rose to the rank of Brigadier, and returned to geology after the war.
- Capstone — A material that resists erosion and protects underlying layers of susceptible material from erosion. Such material forms a cap over a ridge, butte, mesa, or scarp.
- Cementing — A specific process that turns sediment into rock. Water saturated with certain minerals such as calcium sulfate soaks the sediment, precipitating the chemicals as it evaporates. The precipitate fills the pores in the sediment, cementing them into rock.
- Clay-Bearing Unit — This unit on Mount Sharp is a band above the Vera Rubin Ridge and below the Sulfate-Bearing Unit. The CRISM orbital camera shows a high spectrograph value for clay there. Curiosity in-situ data confirms a high value of clay, but not so high as CRISM, relative to surrounding areas.
- Clay — This material's crystal structure forms a two-dimensional sheet of silicon, aluminum, and oxygen. Various other elements are included in the crystals. Stacks of such sheets form the bulk material with some water held within the crystals. Clay has been studied as a possible scaffold for the origin of life. Hardened into rock, it becomes shale.
- Contact — The line or surface separating a younger unit from an older unit. If erosion has removed a different unit between them or allowed a different unit to form between them, it is called an *unconformity*.
- Crater — A nearly circular depression in a surface, usually due to either a meteor strike or a volcanic vent. International rules discourage the use of "Crater" in a name, preferring Gale crater to preserve the ambiguity in "crater" as a class of shape.

C. J. Byrne, *Travels with Curiosity*, https://doi.org/10.1007/978-3-030-53805-7

- Cross-bedded — When sediment is deposited in hilly forms like dunes, the surfaces that define separate hills are wavy or even intersecting. When such deposits are turned into rock, contacts between separate units are not all nearly horizontal and may intersect. Cross-bedding can occur in rocks formed as sediment by wind or slowly flowing water.
- Crystal — When atoms of a mineral are bound in regular arrays, the mineral is called a crystal. A crystal can be as small as a grain, as large as a gemstone, or even a rock. A material can have crystals of several different minerals as components. If the total of the crystals dominates the amorphous components, it is called *crystalline*.
- Delta — A river that flows into a large body of water and deposits much of the sediment it has picked up in its flow as it is no longer in turbulent motion. The settled material builds up in flood conditions, forming levees along the banks and also building up islands and dams that cause the river to split into tributaries. The resulting body of land and streams has the shape of a Greek letter D, hence the term *delta*.
- Diagenesis — When a unit of rock has taken a primary form, a change in the environment can cause elements in its composition to be removed and redeposited in a different unit of rock. For example, if flowing water is in contact with the rock, and the pH of the water changes from neutral to acidic, certain elements can be dissolved, changing the chemistry of the minerals in the first rock. The mobilized elements are moved to a second rock with different chemistry, where the mobilized element may form a different mineral. In this case, each unit is said to have undergone a process called *diagenesis*. In other words, each unit has formed and then been formed again. Awareness of this process can influence judgment of the two different environments when the rock units were formed and then re-formed.
- Dike — New material that is fluid, like compounds dissolved in water or flowing lava, can be forced into fractures in older rock units. If the older rock is eroded to leave the younger rock in the fracture above the surface of the older rock, it is called a *dike*.
- Dunes — Wind-blown structures of crystals such as sand. They may be active or not. If they are, an entire dune or field of dunes may have a measurable motion. The height of dunes is widely variable. The shape of dunes depends on the wind velocity and the size and other properties of the crystals. Dunes on Mars are often very similar to dunes on Earth, even in their variety. Some properties of Martian dunes are unique.
- Exposure — A place where an older unit of rock protrudes from a younger unit because the younger unit has eroded away.
- Fluvial — A process where material is transported by a flowing liquid, water, lava, or another liquid like sulfur.
- Formation — A class of rock strata that have a common process of origin. The process can depend on the material, environment, or the relation to other strata. Units of rock that fit the definition of a formation but differ in other attributes are called members and are given unique names and definitions. Members can be differentiated by composition and appearance (*facies*). See *Group*.

- Fracture — Many rocks have fractures, caused by a lithification process such as drying of a fluvial deposit. After forming into rock, fracturing can be caused by tectonic or volcanic forces, heating or cooling, or forces induced by deposit of other strata above them or nearby. Fractures can occur in patterns that are dependent on the composition of the rock. They can be widened by freezing and thawing.
- Graben — A linear recess in a surface that may hold different materials like sand or water transported to a low elevation by wind or water flow. The cause of a graben may be a flow of water or lava.
- Groundwater — Subsurface water can be held in porous rock, soil, and fractures in rocks. The groundwater may seep into higher strata, driven by pressure or capillary forces. Groundwater can produce surface springs and often has dissolved minerals, sometimes at saturation levels. When groundwater comes to the surface, it can evaporate, leaving a class of minerals called *evaporites*.
- Group — A set of formations with common properties. For example, the Mount Sharp group. Groups can be separated into subgroups or combined into supergroups.
- Halo — When groundwater rises by pressure or capillary forces into a crack, it may rise to spread on the surface of the rock, leaving a thin layer of evaporite. If it has not eroded away, measurements of minerals in the evaporite compared to composition in the native rock can allow inference of the original fluid, the rock exposed to the fluid, and the native rock.
- Hematite — A mineral composed of an oxide of iron with a valance of +2. Like magnetite oxide, it is a fairly hard rock and can be a capstone of a ridge or butte, protecting softer rock from erosion. The CRISM spectrographic camera of the Mars Reconnaissance Orbiter reported a stronger spectral peak of hematite on a ridge South of the clay-bearing unit. For a while, it was called the Hematite Ridge, until it was renamed the Vera Rubin Ridge. When Curiosity made in-situ measurements, they found that the hematite content was similar at the top of the ridge to the unit North of the ridge, weakening the capstone assumption.
- Jura — The highest (southernmost) of the three units of the rocks on the top of the Vera Rubin Ridge is considered the Jura member of the Murray Formation. The name "Jura" suggests the geological formation of the Jura Mountains near the northern boundary of Switzerland. That formation is a series of linear ridges that are due to compressive forces resulting from volcanic features in Switzerland, south of the ridges. Although I have not seen any other reference to the speculation, it appears that whoever suggested the name of the Jura member at the top of the Vera Rubin Ridge was thinking that weight of the kilometers of Mount Sharp above the altitude of the Vera Rubin Ridge may have forced up the ridge.
- Linear dunes — These dunes have the form of a series of parallel ridges, an indication of winds blowing at right angles to each other, perhaps at different seasons. The ridge tops are aligned at 45 degrees to the winds.

- Lithification — The process of turning sediment or volcanic ash into rock. There are two types of processes: the combination of heat and pressure or cementing. Heat and pressure produce metamorphic rock. The sediments in Gale crater have been lithified by cementing, a process of minerals such as calcium or magnesium sulfate or iron-bearing compounds being deposited in pores of mudstone or crystalline sediment.
- Magnetite — A crystalline compound of iron where the iron atoms have a valance of +3.
- Member unit — A member unit of a formation has the defining attributes of the formation and additional attributes that are distinct from other members of the same formation. Common unique attributes are composition, grain size, and characteristics of exposed faces of rock, like color, texture, sheen, etc.
- Metamorphic — Heat and pressure can turn one type of rock into another by changing the chemistry of its components. Typically, the new form is denser and harder than the original.
- Mineral — A component of a rock. Its properties include its constituent elements, how they are formed into molecules, and their crystal structure. Some rocks contain grains of several minerals in a matrix of another mineral.
- Mobility — An element or molecule of a rock can be dissolved into flowing water, especially acidic water, and react chemically to modify the composition of another rock. This property is called *mobility*.
- Mons — The international term for a mountain, Aeolus Mons, for example. An astronomer sighted Gale crater and its central peak through a telescope. The peak forms the largest mountain in its region of Mars. The International Astronomical Union has a Working Group for Planetary System Nomenclature, which named the mountain Aeolus Mons for the Greek servant of the gods who had the role of "Keeper of the winds." The IAU names craters on Mars after famous scientists, but not mountains. Their position is supported by the US Geological Survey. The geologists of the MSL Science Team were unaware of this precedent, and following terrestrial culture, named it after a highly respected deceased scientist, Robert P. Sharp, who built the Division of Geological Sciences at Caltech. Today Aeolus Mons and Mount Sharp are alias names, one standard and the other informal. Peer-reviewed papers have been published using each name. Maps sometimes carry both names.
- Mount Sharp group — The Mount Sharp group consists of formations that have been or will be proposed in addition to the Murray and Stimson formations. The Mount Sharp group should refer to the North face of the mountain (formerly named Aeolis Mons) because that is the only area that has been explored in-situ. That does not preclude geologic mapping based on orbital evidence, but the distinction should be noted. Perhaps there should be a separate group, the Aeolis Mons group, for observations based only on orbital data.
- Mudstone — A rock whose dominant component is amorphous—that is, has particles too small be recognized as crystals.
- Murray Formation — The dominant formation of lower and moderate elevations of Mount Sharp; the exposed rocks are lithified sediment. Horizontal faces are bright and fractured. Vertical faces of scarps or displaced rocks are laminated. The dominant

composition is mudstone, with a variable crystalline content. The defining property of the Murray Formation is deposition from slow-flowing water followed by lithification.

- Unconformity — A contact between formations or members that violates the superposition rule that younger units lie directly upon older units. A unit or part of a unit of intermediate age may have eroded away. Alternately, a set of units may be overturned by compressional force.
- Sandstone — A type of rock whose dominant components are crystalline.
- Scarp — A cliff or nearly vertical body of rock, especially in contact with a horizontal surface. If the scarp is soft enough to become eroded, its components pile up at is base, together with rubble and sometimes boulders. Such debris is called a *talus slope*. In time, erosion causes the scarp to retreat.
- Sediment — Material that has moved from one place to another, usually eroded from rock and carried by water until it settles out when the flow slows. The settling can occur in a riverbed, a lake, an ocean, or even on flood plains of rivers or streams. Material that is eroded by wind (or volcanic ash) can be moved to dunes or other means, including being settled into flowing or still bodies of water. Sediment can also be precipitated from chemicals dissolved in water.
- Siccar Point — A location in Scotland where two diverse formations of rock, one formed in the depths of the ocean and the other formed in a desert at a different time, are in contact. The concept that the rocks inspired is: just because two formations are in contact does not mean that their modes of formation need be related. On Mars, at Mount Sharp, two groups of geologists are using the term *Siccar Point group* to distinguish between sediment deposited in water from sediment deposited by wind at a much later time. One set of geologists is using mostly orbital camera and spectrometer data and the other uses mostly in-situ data from Curiosity. Both distinguish the Siccar Point group from the Mount Sharp group but have different views at the formation level. As new information from the ongoing investigations grows, more formations and members will be found and the relationships will become clearer. The position taken in this book is that the older Murray formation and the younger Stimson formation are in the Siccar Point group. The Mount Sharp group makes room for new formations that do not have the special relations unique to the Murray and Stimson formations. It is possible that at some future time, the Siccar Point group will be absorbed into the Mount Sharp group, perhaps as a subgroup.
- Silica — The mineral silica is silicon dioxide, usually crystalized. Sand is usually composed largely of silica, combined with other components. Silica is soluble in water and can be transported in groundwater. It may be deposited in fractures of other rocks.
- Stimson Formation — A formation that overlies the Murray Formation at lower elevations of Mount Sharp. The color of Stimson rock is darker than the Murray Formation rocks. The dominant component is crystalline sand, with various secondary components. Near a contact with the Murray Formation rocks, fractures in the older rocks are sometimes filled with silica, indicating groundwater flow from the overlying Stimson Formation. The Stimson Formation is heavily eroded, so much so that the underlying Murray rocks are frequently exposed.

- Strata — The fundamental concept of sedimentary geology. At a particular time in geologic history of a planet, sediment forms nearly horizontal layers over wide regions. As it becomes rock through lithification processes, environmental conditions influence properties of the resultant rock. A *strata* is a body of particular properties such as color, composition or formation process that contrast with rocks that were formed earlier or later.
- Sulfate-bearing unit — The imaging spectrometer CRISM in orbit caught a strong signal of sulfate in a band of very bright material between the clay-bearing unit below it and the highest elevations of Mount Sharp above it. In-situ observations by Curiosity of this unit was selected as a major goal in choosing Gale crater as the landing site among the final four candidates. Based on observations by Curiosity of material brought to lower levels by groundwater and filling veins in fractures of the rock there, the dominant chemical of the sulfate-bearing unit is probably calcium sulfate.
- Unit — A body of rocks with similar characteristics that is under investigation to determine its geologic classification.
- Vallis — International term for a valley.
- Volcanic — One of the two broadest classifications of solid material in the crust of a planet. The alternate is sedimentary. Volcanic material comes partly from below the crust and partly from the crust itself, modified by being in or near a volcano.

Index

C. J. Byrne, *Travels with Curiosity*, https://doi.org/10.1007/978-3-030-53805-7